Torsten Richter | Nabil A. Fouad

Guidelines for Thermography in Architecture and Civil Engineering

Theory, Application Area, Practical Implementation

1st English Edition

Birkhäuser

Contents

1 Introduction

1.1 Introduction and Problem Statement

In regions with colder climates, buildings have to be heated as the ambient air temperature changes on a seasonal basis. This is necessary both to provide a healthy indoor climate with sufficient warmth for the building's occupants or residents and also to protect the building stock. Within the framework of resource efficiency and reduction of greenhouse gas emission, the requirements for energy-saving thermal protection increase. General objectives of thermal protection can be summarized as follows:

Objectives of legal requirements for thermal protection:

- Healthy and comfortable indoor climate
- Protection of the construction building against moisture damage
- Saving of heating costs
- Reduction of energy consumption for heating, and thus of the environmental burden.

However, experience shows that in many cases these requirements have not been met. As a result, construction faults, defects, and damages occur.

According to the most recent construction damage report by the German government in 1995, the avoidable costs due to these deficiencies in building construction (in Germany) amount to about 3.4 billion euros. [7] A major part of this sum has to be ascribed to insufficient planning and implementation procedures with regard to thermal insulation.

By means of innovative, nondestructive, and quickly applicable measuring and analyzing methods such as construction thermography, the foundation is laid to localize such deficiencies and initiate further measures to eliminate them. In this context, thermography is to be considered as a tool; only expertise and experience can provide a comprehensive evaluation to a given situation.

This book is aimed at bringing, in condensed form, the existing experiences with regards to the implementation of passive thermography in civil engineering closer to the reader and thus show the possibilities of the technique – but also its realistic limits.

2   Physical Basics
of Thermography

2.1   Basic Principle

Thermography is based on the fact that a body with a temperature above absolute zero (–273.15°C) emits its own characteristic radiation. This phenomenon is caused by the inner mechanical movement of the molecules which is present in any body. The intensity of the molecules' movement depends on the temperature of the considered body. Since charge movement and molecule movement are equipollent, the body emits energy in the form of electromagnetic radiation. This form of radiation moves at the speed of light and obeys the laws of optics. Thus, electromagnetic radiation can be deflected, bundled by means of lens systems or reflected by surfaces. Thermal radiation is in the infrared wavelength range and thus is not visible to the human eye (Fig. 2-1).[1]

Infrared radiation has a wavelength range of 0.78 μm to about 1000 μm. However, for reasons later explained (see Fig. 2-18), only the wavelength ranges of 3 μm up to 5 μm and 8 μm up to 14 μm are of particular interest for thermography in civil engineering.

The objective of thermal imaging is to measure the intensity of the emitted characteristic radiation and thereby to determine the temperature of the emitting (thermographed) surface or body without contact.

Since thermography is closely linked with radiometry, relevant laws of radiation will be explained below.

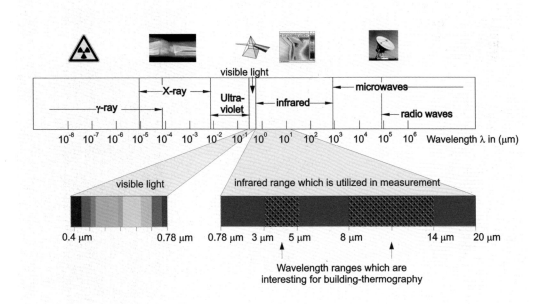

Fig. 2-1 Electromagnetic spectrum.

---

1 *infra*: Latin for "under, below," infrared indicates that infrared light is optically ranged below red light. The denomination refers to the wavelength, which is higher than that of visible light.

2.1.1 Law for Black-Body Radiation

In order to understand the principles of radiation of a real body, it is useful to first represent the laws of radiation using an idealized body. In physics, this type of body is known as the *ideal black body* and is defined as the body with the highest intensity of emitted radiation out of all bodies with the same temperature.

A practical realization of a black body could be a soot-blackened, hollow cube with a little cylindrical opening and walls that are impervious to infrared light. The body is in a state of thermal equilibrium, which means that the temperature of the walls is constant and doesn't change over time. Thus, the exchange of radiation only takes place through the little opening in the cube. Since the temperature of the black body remains constant, the radiation coming in through the opening and being absorbed by the soot-blackened interior surface has to be equal to the emitted radiation.

According to the laws of Josef Stefan (1879, experimental discovery) and Ludwig Boltzmann (1884, physical derivation) the specific radiation M, which is emitted from the black body through the opening in the cube, depends only on the inner temperature of the black body and is calculated as follows:

$$M \, [W/m^2] = \sigma \cdot T^4 = C_s \cdot (T/100)^4 \tag{1}$$

with:

σ Stefan-Boltzmann constant = $5.67 \cdot 10^{-8}$ W/(m²K⁴)

C_S Coefficient of radiation of black body = 5.67 W/(m²K⁴)

T Absolute temperature T [K] = $273.15 + \theta$ [°C]

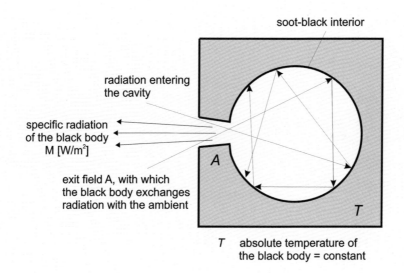

Fig. 2-2 Model of a black body.

The specific radiation M given in equation (1) represents the sum of energy over the whole wavelength spectrum. Therefore, the distribution of black-body radiation over the wavelength is of special interest. Planck's radiation law (Max Planck, 1900) describes the correlation between the temperature of the black body and the specific radiation depending on the wavelength. The spectral specific radiation according to Planck's radiation law is given below:

$$M_\lambda\,[W/(cm^2 \cdot \mu m)] = \frac{c_1}{\lambda^5} \cdot \frac{1}{\exp\left(\dfrac{c_2}{\lambda \cdot T}\right) - 1} \tag{2}$$

with:

c_1 1st radiation coefficient = $3.7418 \cdot 10^4\,[W \cdot cm^{-2} \cdot \mu m^4]$
c_2 2nd radiation coefficient = $1.4388 \cdot 10^4\,[K \cdot \mu m]$
T Absolute temperature $T\,[K] = 273.15 + \theta\,[°C]$
λ Wavelength $[\mu m]$

Fig. 2-3 shows Planck's radiation law for varying absolute radiation temperatures. Furthermore, the diagram depicts the wavelength visible to the human eye. It is clear that only hot bodies radiate in a wavelength range that is visible to the human eye.

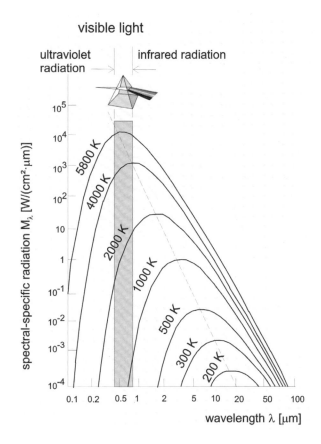

Fig. 2-3 Planck's radiation law: Characteristics of a black body, depending on temperature and wavelength; the shaded area depicts the range visible to the human eye.

According to Planck's radiation law all curves exhibit the same shape, yet they never cross one another. The areas under the curves represent the total amount of energy a black body emits under a certain temperature. Integrating equation (2) over the wavelength leads to the Stefan-Boltzmann law, which has already been mentioned in equation (1).

$$M = \sigma \cdot T^4 = \int_{\lambda_1}^{\lambda_2} M_\lambda \, d\lambda \tag{3}$$

Fig. 2-3 also shows that with increasing absolute temperature the specific radiation maximum (dotted line) shifts to shorter wavelengths. The location of the specific radiation maximum; the wavelength at which the body exhibits the maximum radiation, can be calculated by means of Wien's displacement law (Wilhelm Wien, 1893):

$$\lambda_{max} = \frac{2898 \, \text{K} \cdot \mu\text{m}}{T} \tag{4}$$

with:
λ_{max} Wavelength at radiation maximum [μm]
T   Absolute temperature T [K] = 273.15 + θ [°C]

Tab. 2-1 lists radiation maxima calculated by means of Wien's law for typical radiation temperatures:

Tab. 2-1 Specific radiation maxima depending on the absolute radiation temperature.

Body	Absolute temperature [in K]	Wavelength λ_{max} at max specific radiation [in μm]
Surface temperature of the sun	5800	0.50
Iron, brightly glowing	1200	2.42
Water at boiling point, 100°C	373	7.77
Human body temperature, 37°C	310	9.35
Exterior wall (Inside), 17°C	290	9.99
Freezing point of water, 0°C	273	10.62
Exterior wall, −5°C	268	10.81

The distribution of the specific radiation shown in Fig. 2-3 can be computed and is listed in Tab. 2-2 depending on the radiation density maximum; see equation (4):

Physical Basics of Thermography

Tab. 2-2 Percentages of the total radiation depending on the radiation maximum (λ_{max}); for calculation see [41].

Wavelength range λ_{max} [μm]	Rounded percentage of total radiation [%]
(0 to 0.5) · λ_{max}	1
(0 to 0.7) · λ_{max}	5
(0 to 0.8) · λ_{max}	10
(0 to 0.9) · λ_{max}	20
(0 to 1.0) · λ_{max}	25
(0 to 1.1) · λ_{max}	30
(0 to 1.2) · λ_{max}	40
(0 to 1.4) · λ_{max}	50
(0 to 1.7) · λ_{max}	60
(0 to 1.9) · λ_{max}	70
(0 to 2.4) · λ_{max}	80
(0 to 3.3) · λ_{max}	90
(0 to 4.3) · λ_{max}	95

Explanation Tab. 2-2:

The maximum radiation density of a warm wall with a surface temperature of 17°C (290 K) is about $\lambda_{max} \approx 10$ μm (see Tab. 2-1). The IR camera system mainly used for the thermograms in Chapters 7 and 8 has a receiver detector which can only process IR radiation within a range of 8–14 μm.

For the thermograms that will be discussed in Chapters 7 and 8 later on, an IR camera system contained a receiver detector which can only process IR radiation within a range of 8–14 μm. Ten percent of the total radiation is emitted within the spectral range from 0 to 0.8 · λ_{max} (from 0 to 8 μm); about 50% is emitted within the spectral range from 0 to 1.4 · λ_{max} (from 0 to 14 μm). Therefore, even if the radiation is completely absorbed (no losses), the detector of the abovementioned IR camera could also only detect a maximum of 40% (50%–10%) of the total emitted radiation. Due to losses in the passage of the radiation through the lenses, the actual radiation arriving at the detector is even lower.

The following three explanations offer a better understanding of this statement:

- When heating steel, a dark-red glow of the steel is observed from a temperature of about 650°C (923 K). This is about the temperature at which the radiation characteristics reach the range visible to the human eye. However, the radiation maximum is not part of the visible range, but is located at about 3.1 μm, according to Wien's displacement law – in the infrared range.
- Due to the high surface temperature of the sun (T ≈ 5800 K), a high amount of radiation is emitted. The radiation maximum is in the range visible to the human eye. Thus, solar radiation can easily be perceived by the human eye.
- The temperature for building construction thermography usually ranges from −20°C to +20°C. The radiation maximum reaches from about 9.8 μm to 11.4 μm, which is within the infrared range. The hardware for construction thermography (IR camera) is optimized for this particular range.

2.1.2 Law for Real Bodies Radiation

When electromagnetic radiation hits a real body, parts of the radiation are absorbed (transformed into heat), reflected (thrown back into space) or, in case of permeable bodies, transmitted. The amount of absorbed, reflected, or transmitted radiation depends on the properties of the considered body and the wavelength of the incident radiation. The energy balance is as follows:

Incident radiation: $M = A + R + T$ (5)

with:
M   incident radiation
A   absorbed radiation
R   reflected radiation
T   transmitted radiation

With the calculation step (divided by M), equation (5):

$M = A + R + T$ $| : M$

leads to

$$1 = \frac{A}{M} + \frac{R}{M} + \frac{T}{M} = \alpha(\lambda) + \rho(\lambda) + \tau(\lambda) = 1 \tag{6}$$

with:
α   absorption coefficient, function of the wavelength λ
ρ   reflectance, function of the wavelength λ
τ   transmittance of the body, function of the wavelength λ

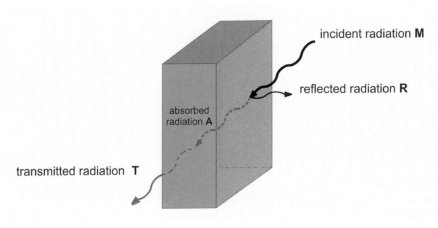

Fig. 2-4 Possible split of incident radiation.

Physical Basics of Thermography

Idealized limit cases are:

Tab. 2-3 Idealized limit cases of radiation bodies.

Limiting case	Description	Parameters
ideal black body (nontransparent body)	incident radiation is fully absorbed	Absorption coefficient $\alpha = 1$ Reflectance $\rho = 0$ Transmittance $\tau = 0$
ideal mirror (nontransparent body)	incident radiation is fully reflected	Absorption coefficient $\alpha = 0$ Reflectance $\rho = 1$ Transmittance $\tau = 0$
ideal window (transparent body)	incident radiation is fully transmitted	Absorption coefficient $\alpha = 0$ Reflectance $\rho = 0$ Transmittance $\tau = 1$

In the previous section, radiation properties of the black body were discussed. Characteristic for a black body is the complete absorption of the incident radiation. According to Kirchhoff's radiation law (Gustav Kirchhoff, 1859), this body emits the same amount of radiation since it doesn't heat up by definition.

However, real bodies always emit less radiation than the idealized black body. The link between the model of the black body and real objects is established by means of Kirchhoff's radiation law. According to that law, the radiation capability (that is, emissivity) of a random body divided by the emissivity of a black body is equal to the ratio of the absorption coefficients of both bodies.

It is:

$$\varepsilon(\lambda,T) = \frac{M_{real}}{M_{black\,body}} = \frac{\alpha}{\alpha_s} \tag{7}$$

with:

ε emissivity of real body
M_{real} radiation emission of a real body
$M_{black\,body}$ radiation emission of a black body
α absorption coefficient of the real body
α_s absorption coefficient of the black body

Because $\alpha_s = 1$ (see Tab. 2-3, black body):

$$\varepsilon = \alpha \tag{8}$$

This means that according to Kirchhoff's law, the absorption coefficient α (λ) of a body in thermal equilibrium is equal to its emissivity ε (λ), where the absolute magnitude of absorption coefficient and emissivity are dependent on the present wavelength range. Furthermore, the following parameters may also have a certain impact:

- texture of the surface
- material composition
- temperature
- angle to reflected beam.

Equation (6) can also be written as follows:

$$\varepsilon(\lambda) + \rho(\lambda) + \tau(\lambda) = 1 \qquad (9)$$

with:

ε emissivity, function of the wavelength λ

ρ reflectance, function of the wavelength λ

τ transmittance, function of the wavelength λ

Fig. 2-5 shows a curve of emissivity ε for the three limited cases (black body, selective radiator, gray body) depending on the wavelength.

For the special case of a black body, the emissivity $\varepsilon = 1$; for an ideal mirror and an ideal window, $\varepsilon = 0$ (see Tab. 2-3). In the case where the emissivity is constant over a wavelength range, the object looked at is called a *gray body*. If the emissivity exhibits enormous changes throughout the wavelength, it is called a *selective radiator*.

Needless to say, because of the different emissivity, the Planck's radiation curve (specific radiation) of real bodies has to change as well. It is:

$$M = \int_{\lambda_1}^{\lambda_2} \varepsilon \cdot M_\lambda \, d\lambda \qquad (10)$$

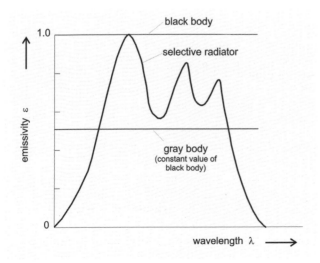

Fig. 2-5 Dependency of the emissivity on the wavelength.

Physical Basics of Thermography

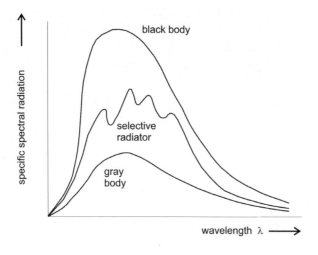

Fig. 2-6 Characteristic dependency of specific radiation on emissivity.

Emissivity ε of Real Surfaces

Within the wavelength range (of longwave heat radiation) interesting for construction thermography, many nonmetallic, nonshiny materials exhibit a very high and quite constant emissivity (ε = 0.80–0.95). For construction thermography, those materials can be assumed to be so-called gray radiators. However, the actual color of the object is not of interest (see Fig. 2-7).

Metals with polished or shiny surfaces especially reveal a considerable reflection of the radiation within the abovementioned wavelength range. Thus, the absorption coefficients as well as the emissivity are very low. Therefore, materials such as silver and gold – but also polished aluminum – are very well suited for mirror material.

The general behavior of the emissivity of various materials over different wavelengths is shown in Fig. 2-7.

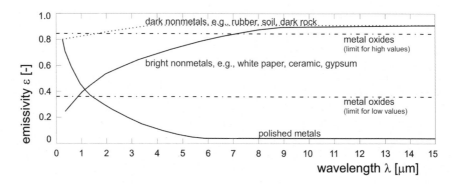

Fig. 2-7 Emissivity of various materials at room temperature in dependence of the wavelength; for further information see [41].

Knowing the correlations depicted in Fig. 2-7, it is easy to understand why it doesn't make a difference if a radiator with a temperature of 40°C to 70°C is painted black or white, because the heat radiation is primarily emitted in the longwave range (> 8 μm). Within this wavelength range the emissivity of bright and dark nonmetals do not differ.

Emissivity is usually determined by means of laboratory testing. Approximated values in dependency of temperature and wavelength range can also be drawn from tables (for example, from [25] or [39]). Moreover, the emissivity has to be adjusted to the wavelength range of the used detector of the IR camera. Tab. 2-4 lists a selection of emissivity and their scope of application:

Tab. 2-4 Emissivity of various surfaces.

Materials and texture of surface in general for temperatures interesting for construction thermography (−10°C to 100°C), Range wavelength: λ = 8–14 μm	Emissivity ε
Metals, high-gloss polished	
Aluminum, polished	0.04–0.06
Silver, polished	0.02–0.05
Metals in technical condition	
Aluminum, raw	0.06–0.07
Aluminum, metallic	0.02–0.15
Aluminum, heavily oxidized	0.6–0.8
Iron, steel, completely rusted	0.6–0.8
Iron, steel, galvanized	0.2–0.3
Copper, oxidized	0.2–0.7
Brass, polished	0.2–0.3
Coating	
Aluminum finish	0.1–0.15
Enamel varnish, snow-white	0.7–0.90
Other materials	
Concrete	0.90–0.95
Water, snow, ice	0.80–0.95
Glass, smooth	0.80–0.95
Marble, bright gray, polished	0.90–0.95
Human skin	0.95–0.98
Rocks, soil	0.60–0.95
Timber	0.90–0.95
Cloth	0.75–0.95
Paper	0.70–0.95

Physical Basics of Thermography

At this point it has to be reemphasized that the emissivity is equal to the absorption coefficient only for bodies in thermal equilibrium. A brick, for example, with a temperature of 20°C receives a longwave radiation of a 20°C radiation source with an absorption coefficient of $\alpha \approx 0.95$. Since the brick itself also has a temperature of 20°C, it emits heat radiation with an emissivity of $\varepsilon \approx 0.95$.

However, if the brick with a surface temperature of 20°C is hit by shortwave solar radiation (radiation temperature of about 5500°C), the radiation is absorbed by an absorption rate of $\alpha \approx 0.50$ (shortwave emissivity). Due to solar radiation, the temperature of the brick will rise a few Kelvin to, for example, 25°C. Because of the relatively low temperature (25°C) the brick only emits longwave radiation with an emissivity of $\varepsilon \approx 0.95$ (see Tab. 2-5 and Fig. 2-7).

Tab. 2-5 Emissivity of various surfaces depending on the radiation wavelength.

	Longwave heat radiation (emitted by bodies with a temperature ≈ 20°C) emissivity ε = absorption coefficient α	Solar radiation = shortwave heat radiation (emitted by bodies with a temperature ≈ 5500°C) emissivity ε = absorption coefficient α
Brick	approx. 0.95	approx. 0.50
Concrete (smooth)	approx. 0.95	approx. 0.50
Window glass	approx. 0.90	0.04 to 0.40 (Depending on transmittance)

A good way to show the impact of different emissivity values is the Leslie cube (named after its inventor, physicist John Leslie, 1804). The Leslie cube is a hollow brass cube, which can be filled with hot water so that all sides of the cube exhibit the same surface temperature. The four sides of the cube have different coatings with different emissivity (matte metallic, polished metallic, white coating, black coating).

Fig. 2-8 Leslie cube with description of the surface properties. The cube is filled with hot water (about 50°C), the walls of the cube thus all have the same surface temperature.

In general, thermograms are generated with one emissivity valid for the whole picture (for example, here ε = 0.98). In this case the emissivity is suitable for the black and white surfaces of the cube and the temperatures of those surfaces are represented correctly. However, the metallic surfaces exhibit a considerably lower emissivity. Physically speaking, metallic surfaces emit far less heat, which is represented by the low emissivity; see equation (3).

Thus, the temperature, which is about 48°C in real terms on all sides and also on the polished metallic side, is not correctly displayed due to the choice of an emissivity that is too high (display value only 23°C, see Fig. 2-9).

Date	Temperature boundary conditions		Additional information
	Exterior temperature	Interior temperature	
07.01.2008	approx. 22°C (room temperature)	approx. 48°C (temperature inside the cube)	The hot water is filled to about 1 cm below the top of the cube (this is why the top is cooler).

 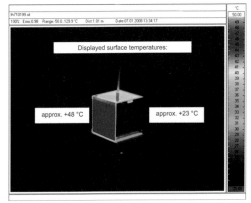

Fig. 2-9 Thermogram of a Leslie cube with a chosen global emissivity in order to illustrate the impact of different actual emissivity values of the surfaces. The surfaces of the cube have a steady temperature of about 48°C, but in the thermogram the temperature of the right side of the cube (polished metallic) is significantly lower than the actual temperature.

When using thermography in civil engineering, metallic features or window coatings can lead to false displayed temperatures due to their very different emissivity. A practical example of this problem is shown below: The thermogram reveals a significantly lower temperature along the heating pipe, but there is no temperature difference in reality. The problem is that the global emissivity is not suitable for the uncoated section of the pipe (see Fig. 2-10).

Besides the IR radiation emitted by the object itself, some surfaces exhibit a directed reflection (Fig. 2-11). Since the IR radiation follows the laws of optics, the ambient radiation is reflected by very smooth and planar surfaces in the thermogram.

Instead of only measuring the emitted radiation of the measurement object, the temperature of the reflected object is analyzed. This effect can be seen when thermographing

Physical Basics of Thermography

Date	Temperature boundary conditions		Additional information
	Exterior temperature	Interior temperature	
02.07.2009	./.	approx. 20°C	./.

 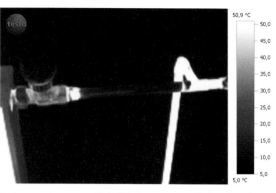

Fig. 2-10 Choosing a global emissivity doesn't represent the temperature correctly in this thermogram of a heating pipe due to the different real emissivity of the black- and white-coated pipe sections.

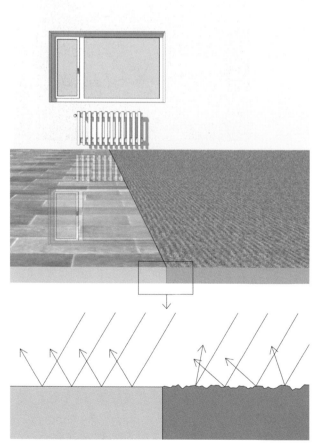

Fig. 2-11 Explanation of directed reflection: on the left side, the directed reflection on the smooth floor (tiles) reproduces an image on the surface; on the right side, the reflected radiation is diffuse (reflects in all directions), and no image is reproduced.

window areas or smooth tiling (see Figs. 2-12 and 2-13). Especially when thermographing exterior window areas from below, the IR camera detects the sky reflected in the windows. This phenomenon is clearly noticeable on thermograms when the sky is clear and cold. Directed reflection also plays a role when thermographing indoors – for example, on smooth floors – as mentioned above. When thermographing windows from indoors, the person conducting the measurement can oftentimes see himself in the reflected image. Because of its characteristics, infrared radiation can also be reflected on surfaces that do not reflect visible light (see Fig. 2-13).

Normally reflections are easy to detect and/or to eliminate, when the object is thermographed from various angles and changing shadows, or so-called hot spots, are observed due to a change in thermographing location. To do this, move the camera while thermographing and follow the changes of the hot spot in the thermogram. If the hot spot moves synchronously with the camera movement and appears on several areas of the surface, a reflection can be assumed. Thereby it is easy to detect thermographs influenced by reflections while measuring. However, it's much more difficult to identify reflections in the picture afterward.

Fig. 2-12 Reflections on a window surface: The thermogram doesn't show the actual surface temperature of the window but rather the temperature of the atmosphere.

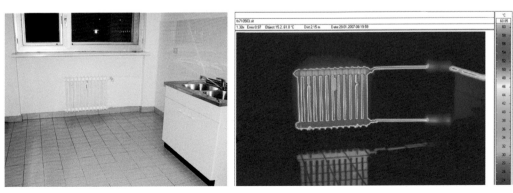

Fig. 2-13 Reflection of a hot radiator on tiling: As can be seen in this picture, reflections can occur even on highly emitting surfaces; the texture of the surface is decisive.

Dependency of the Emissivity on the Angle of Observation

The emissivities given in reference books are usually for a frontal (vertical) view. If the perspective changes from perpendicular (observation angle = 0°) to steeper angles (max. observation angle = 90°), the resulting emissivity changes too. Such a situation is easily illustrated during the thermography of high-rise buildings, since the image is usually produced close to the building and the upper floors can only be viewed at a steep angle.

Fig. 2-14 Steep angles often occur in thermograms of tall buildings. By changing the emissivity from angles of approx. 60°, emissivity corrections may become necessary.

For nonconductive materials, the emissivity decreases for observation angles of view larger than α = 60°. Metallic surfaces normally exhibit very small emissivity for small polar angles. Interesting however, is a significant maximum for an angle of view of about α = 85°. At that angle, emissivity values of ε > 0.4 are not considerably exceeded (see Fig. 2-15).

The varying emissivity values are of special interest, since they lead to an increase of reflectance. Under the assumption of τ = 0, which is reasonable for opaque objects, and transposition of equation (9), the reflectance ρ_M can be written as follows:

$$\varepsilon_M + \rho_M = 1 \quad \text{converted:} \quad \rho_M = 1 - \varepsilon_M$$

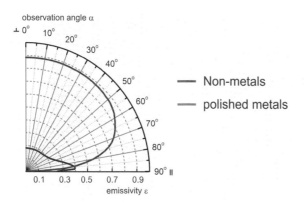

observation angle α

— Non-metals

— polished metals

emissivity ε

Fig. 2-15 Dependency of the emissivity on the perspective (polar angle α), figure taken from [4].

Therefore, the change of the emissivity ε and the reflectance ρ$_M$ have to be taken into account when thermographing an object from an angle other than perpendicular. Surfaces being thermographed from a tilted angle, such as facades or bent parts of furniture, often exhibit significant reflection phenomena due to the optical path of radiation that creates a slanting striped surface.

By means of the abovementioned Leslie cube (Fig. 2-9), the practical impact of a changed emissivity due to the observation angle can be illustrated. The Leslie cube sits on a revolving mount; the black side is thermographed from different perspectives. The depicted temperatures are documented and the emissivity that shows the correct temperature is displayed (the real surface temperature is always the same, approx. 55°C). The emissivity exhibits the typical behavior of the graphs shown in Fig. 2-15 and Fig. 2-16.

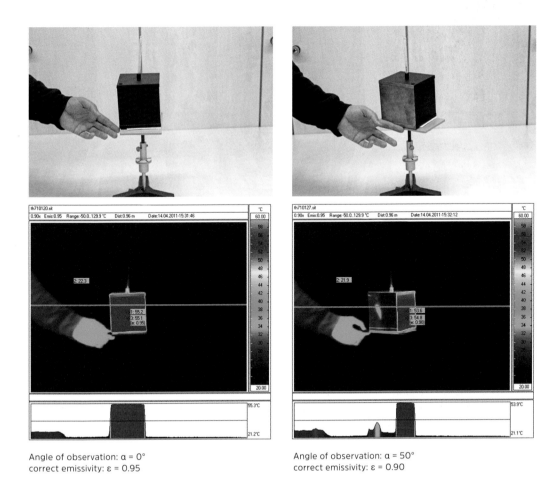

Angle of observation: α = 0°
correct emissivity: ε = 0.95

Angle of observation: α = 50°
correct emissivity: ε = 0.90

Fig. 2-16 – Part 1 Illustration of the impact of different observation angles on the temperature displayed: For retrograde calculation of the actual, real temperature, the correct emissivity is given.

Physical Basics of Thermography

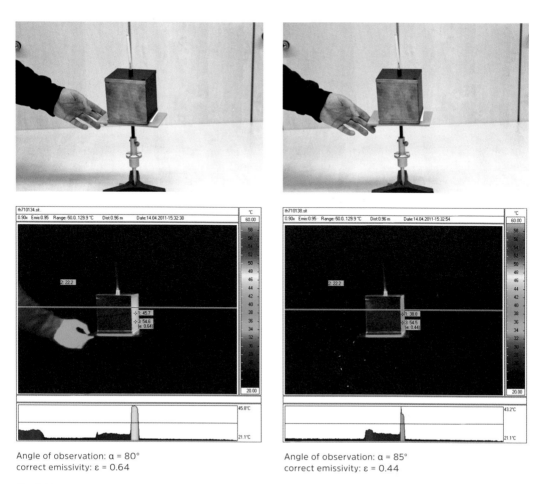

Angle of observation: α = 80°
correct emissivity: ε = 0.64

Angle of observation: α = 85°
correct emissivity: ε = 0.44

Fig. 2-16 – Part 2 Illustration of the impact of different observation angles on the temperature displayed: For retrograde calculation of the actual, real temperature, the correct emissivity is given.

Behavior of Glass in Thermography

Glass is an important construction material which is often thermographed because of its use in windows. At this point it must be mentioned that glass is pervious to shortwave radiation but not to longwave heat radiation (see Fig. 2-17).

This fact is also the reason for the so-called greenhouse effect in extensively glazed buildings: shortwave radiation transmits through the glass, hits the inner construction surface and is transformed into heat. The warm surface now emits longwave heat radiation that is blocked by the glass (important for protection against summer overheating and for passive usage of solar energy). The radiative properties of glass are illustrated in Fig. 2-18. It is clear to see that the transmission rate τ is very high in the visible ranges and the bordering infrared range, which means that it is transparent in these ranges. According to the physical laws the emissivity ε is small in these sections; see equation (9). The diagram shows a steep increase

Fig. 2-17 Normal window glass is pretty much impervious to heat radiation; you can also see the thermographic camera mirrored in the image.

of the emissivity ε to more steady values around ε = 1 from a wavelength of 4.5 μm on. The transmittance τ decreases accordingly. In the further wavelength range the emissivity exhibits larger irregularities.

While glazing is rather impervious for infrared radiation, thin plastic films may show high transmittances for infrared radiation. Construction sites provided with screens for visual protection could under certain circumstances also be thermographed. In most cases, the evaluation based on the experience of the human eye doesn't apply for infrared radiation (see Fig. 2-19).

Summary:

- The color of the surface is negligible with regards to absorption and emission for longwave radiation (heat radiation). For longwave radiation transfer, a distinction has to be made between metallic surfaces (ε ≈ 0.05–0.2) and nonmetallic surfaces (ε ≈ 0.90–0.98).

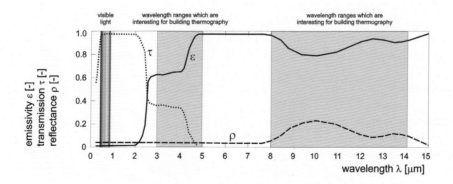

Fig. 2-18 Emissivity, transmittance, and reflectance of glass depend on the wavelength [μm]; for every wavelength λ it is: ε + τ + ρ = 1.

Physical Basics of Thermography

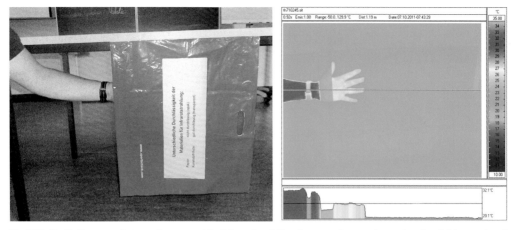

Fig. 2-19 Plastic films are often very transparent for infrared radiation; however, they are impervious for visible wavelengths.

- For shortwave radiation (such as solar radiation), the color of the surface plays an important role (see dark/bright nonmetals with wavelengths < 8 µm in Fig. 2-7). Dark surfaces absorb shortwave radiation better than bright surfaces and therefore heat up more quickly.
- Emissivity is strongly dependent on the observation angle. If possible, the thermogram should be taken perpendicular to the thermographed object.
- In general, glass is impervious to longwave heat radiation but transparent for shortwave radiation.
- Plastic films are transparent for infrared radiation. Therefore, the surface temperature of plastic films depicted in thermograms is often influenced by the transmitted radiation.

2.2 Effects of the Environment on Measuring

2.2.1 Effects of the Atmosphere

Infrared thermography is a noncontact measurement, which means the energy emitted by a body first has to penetrate the atmosphere before it can be detected. Penetrating the atmosphere, the emitted energy is absorbed, reflected, and scattered by the particles in the air. Thus, the further away the receiver is from the measuring object, the less radiation can be detected.

Studies have shown that radiation is weakened mainly by three overlapping reasons:

- molecular absorption by humidity (H_2O)
- molecular absorption by carbon dioxide (CO_2)
- scattering by molecules and particulate matter.

This so-called transmission behavior of the transmission distance is wavelength-dependent. In some wavelength ranges, the atmosphere exhibits a high transmittance; in others, a rather low one.

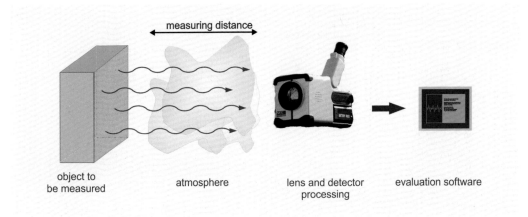

measuring distance

object to
be measured

atmosphere

lens and detector
processing

evaluation software

Fig. 2-20 Principles of noncontact thermography.

Fig. 2-21 shows a schematic illustration of the transmission behavior of carbon dioxide and humidity (water vapor) in dependency of the wavelength. The amount of CO_2 in the air is quite constant and so is its transmittance. However, since air humidity changes often, the transmittance is highly dependent on the present air humidity (amount of water vapor).

Of course, transmission is also dependent on the distance covered. The further away the source of radiation is from the detector of the IR camera, the higher the damping (lower transmittance). Comparatively, Figs. 2-22 and 2-23 illustrate this correlation.

Furthermore, Figs. 2-22 and 2-23 reveal that there are wavelength ranges within the infrared spectrum in which the transmission is constantly very high without any significant absorption property. These areas are called *atmospheric windows* and mark the sections in which emitted radiation faces very little damping on its way through the atmosphere. By chance,

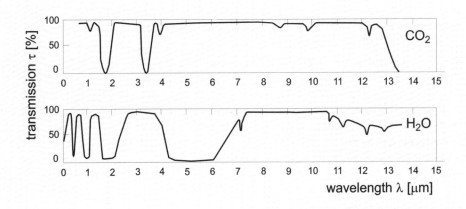

Fig. 2-21 Schematic illustration of the transmittance for CO_2 and H_2O in the form of air humidity for a measuring distance of about 300 m, taken from [41].

Physical Basics of Thermography

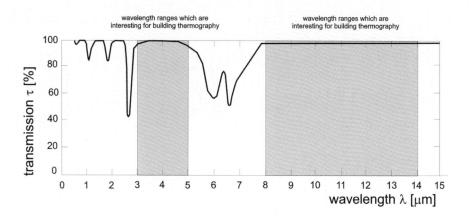

Fig. 2-22 Schematic illustration of the transmittance of a 1 m measuring distance, with an air temperature of 32°C and a relative humidity of 75%.

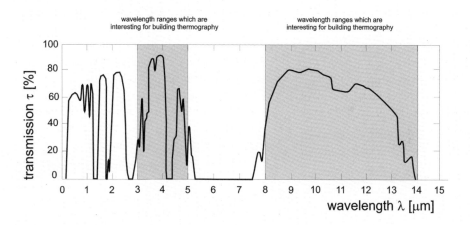

Fig. 2-23 Schematic illustration of the transmittance of a 2000 m measuring distance, with an air temperature of 15°C and a relative humidity of 75%.

one of these atmospheric windows covers the wavelength range of 8–14 µm, which is the maximum of radiation emitted by bodies in the temperature range of interest for construction thermography (see Tab. 2-1). Therefore IR cameras used for construction thermography are optimized for this wavelength range. Another atmospheric window is located in the area between 3 µm and 5 µm.

This optic window coincides with the radiation maxima of higher temperature ranges, as can be observed in casting houses or steelworks, where thermography is primarily used for process monitoring.

Moreover, Figs. 2-22 and 2-23 show that the damping of radiation by the atmosphere is negligible for measuring distances covered in construction thermography (about 3 m for indoors thermography up to about 20 m for outdoors thermography). If more detailed results are demanded, professional IR systems allow a user to consider the influence of atmospheric damping by means of camera-specific evaluation software.

When there is dust or water in the atmosphere (fog, mist, or rain), the scattering of radiation increases considerably. Thus, thermograms taken in bad weather conditions (fog, snow, dusty atmosphere) must be assessed to determine if the results are meaningful (see example in Chapter 8.2.1).

2.2.2 Effects of Ambient Radiation and Background Radiation: Evaluation Equation of Thermography

The infrared detector of a noncontact measurement receives the radiant energy of the measuring object and converts the value into a temperature. In order to display the correct surface temperature, the camera must first be calibrated on an ideal black body.[2]

For a real temperature measurement, the laboratory conditions of a black radiator are not given. As depicted in Fig. 2-24, different shares of the radiation influence the radiation signal that is finally detected by the sensor.

By balancing the shares of radiation, the thermographic basic equation, which is used to convert the detected radiation into a temperature, can be set up. After transposing equation (7), a measured object which has a temperature θ_M and an emissivity ε_M emits the following radiation M_M:

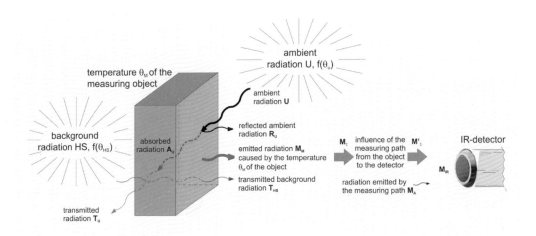

Fig. 2-24 Schematic illustration of the shares of radiation in a thermographic measurement setup.

---

2 Calibration is done by the manufacturer and is usually conducted by means of a black radiator (see the principle illustrated in Fig. 2-2).

$$M_M = \varepsilon_M \cdot M_{black\ body}(\theta_M) \qquad \text{(11)}$$

with:

ε_M emissivity of the measured object

M_M radiation emission of real body at θ_M

$M_{black\ body}(\theta_M)$ radiation emission of black body at θ_M

The measuring object receives ambient radiation that is reflected, absorbed, or transmitted. The part of the reflected ambient radiation that is directed toward the IR sensor can be written as follows:

$$R_U = \rho_M \cdot U(\theta_U) \qquad \text{(12)}$$

with:

R_U reflected share of ambient radiation

ρ_M reflectance of measuring object

$U(\theta_U)$ ambient radiation due to present average temperature θU of half-space of the object

Objects that are transparent for infrared radiation (such as plastic films, gases) let radiation from the background pass through. The intensity of this radiation depends on the radiation temperature of the background (θ_{HS}) and the transmittance of the object.

$$T_{HS} = \tau_M \cdot HS(\theta_{HS}) \qquad \text{(13)}$$

with:

T_{HS} transmitted share of radiation from the background

τ_M transmittance of real body

$HS(\theta_{HS})$ ambient radiation due to radiation temperature θ_{HS} of the background

The shares given in equations (11), (12) and (13) can be summed up in the total radiation M_Σ.

$$M_\Sigma = M_M + R_U + T_{HS} \qquad \text{(14)}$$

In general, the objects analyzed by means of construction thermography can be considered as nontransparent ($\tau_M = 0$) to infrared radiation. (**Caution:** This does not apply for thin plastic films, gases and other transparent materials! See Fig. 2-19.) In the case of nontransparent materials, $T_{HS} = 0$, and equation (9) can be written as follows:

$$\varepsilon_M + \rho_M + \tau_M = \varepsilon_M + \rho_M + 0 = \varepsilon_M + \rho_M = 1 \tag{15}$$

with:

ε_M emissivity of measuring object

ρ_M reflectance of measuring object

τ_M transmittance of measuring object

By transposing equation (15), the reflectance ρ_M of the measuring object can also be written as follows:

$$\varepsilon_M + \rho_M = 1 \quad \text{converted: } \rho_M = 1 - \varepsilon_M \tag{16}$$

with:

ε_M emissivity of measuring object

ρ_M reflectance of measuring object

Now the total radiation for measuring objects that are nontransparent to infrared radiation can be given as follows:

$$M_\Sigma = \underbrace{\varepsilon_M \cdot M_{\text{black body}}(\theta_M)}_{M_M} + \underbrace{(1 - \varepsilon_M) \cdot U(\theta_U)}_{R_U} + \underbrace{0}_{T_{HS}} \tag{17}$$

On its way through the atmosphere (through the measuring path), the radiation M_Σ described by equation (17) is damped depending on the transmittance of the atmosphere (see Section 2.2.1):

$$M'_\Sigma = \tau_{MP} \cdot M_\Sigma \tag{18}$$

with:

M'_Σ damped radiation after passing through the atmosphere

τ_{MP} transmittance of the measuring path (atmosphere)

$M\Sigma$ radiation in the immediate environment of the (real) body (that is, the measuring object)

On the way through the atmosphere, the measuring path with the radiation temperature θ_{MP} can also emit radiation (M_{MP}), which is added to the damped radiation M'_Σ.

Physical Basics of Thermography

Due to the assumption that the measuring path holds a reflectance of $\rho_{MP} = 0$, the emissivity of the measuring path can be written as:

$$\varepsilon_{MP} + \underbrace{\rho_{MP}}_{=0} + \tau_{MP} = 1 \quad \text{converted: } \varepsilon_{MP} = 1 - \tau_{MP} \tag{19}$$

with:

ε_{MP} emissivity of the measuring path
τ_{MP} transmittance of the measuring path

Thus, the radiation emitted by the measuring path can be written as:

$$M_{MP} = \varepsilon_{MP} \cdot M_{black\,body}(\theta_{MP}) = (1 - \tau_{MP}) \cdot M_{black\,body}(\theta_{MP}) \tag{20}$$

with:

M_{MP} radiation due to radiation temperature θ_{MP} of the measuring path
ε_{MP} emissivity of the measuring path
$M_{blackbody}(\theta_{MP})$ radiation emission of a black radiator at θ_{MP}

Through consideration of the equations (17) through (20), the (measurable) radiation at the IR receiver is described by the following equation:

$$M_{IR} = M'_{\Sigma} + M_{MP} = \tau_{MP} \cdot [\varepsilon_M \cdot M_{black\,body}(\theta_M) + (1 - \varepsilon_M) \cdot U(\theta_U)] + (1 - \tau_{MP}) \cdot M_{black\,body}(\theta_{MP}) \tag{21}$$

with:

M_{IR} radiation measured at IR sensor
M'_{Σ} damped radiation after passing through the atmosphere
M_{MP} radiation emitted by the measuring path

In order to display the object temperature that is of interest for thermographic analyses, equation (21) has to be solved for the object temperature θ_M:

$$\theta_M = M^{-1} \cdot \left(\frac{\dfrac{M_{IR} - (1 - \tau_{MP}) \cdot M_{black\,body}(\theta_{MP})}{\tau_{MP}} - (1 - \varepsilon_M) \cdot U(\theta_U)}{\varepsilon_M} \right) \tag{22}$$

Here, M^{-1} describes the camera-specific inverse function (including calibration values) of the IR camera, by means of which the radiation detected by the sensor is converted into a temperature. As can be seen in equation (22), several coefficients of the measuring setup have to be known in order to determine the exact object temperature by noncontact means:

ε_M emissivity of measuring object

θ_U average temperature of half-space surrounding the measuring object to determine ambient radiation

τ_{MP} transmittance of the measuring path

θ_{MP} radiation temperature of the measuring path

In regular circumstances, equation (22) can be simplified, since the atmospheric transmission losses are marginal for short measuring distances and normal humidity (see Section 2.2.1). Furthermore, a marginal influence of the infrared radiation due to the radiation emitted by the measuring path can be assumed. Thus, the equation can be simplified as follows:

$$\theta_M = M^{-1} \cdot \left(\frac{M_{IR} - (1 - \varepsilon_M) \cdot U(\theta_U)}{\varepsilon_M} \right) = M^{-1} \cdot \left(\frac{1}{\varepsilon_M} \cdot [M_{IR} - U(\theta_U)] + U(\theta_U) \right) \tag{23}$$

Equation (23) shows that for an emissivity $\varepsilon_M < 1$ the reflected share of ambient radiation has to be taken into account when computing the temperature of the measuring object (see Fig. 2-24). Furthermore, the impact of ambient radiation or ambient temperature θ_U, respectively, has to be considered when adjusting the emissivity. When indicating absolute temperatures on the basis of thermography, one should always be aware of these facts.

Only in case of an ideal black body ($\varepsilon_M = 1$) does the dependency shown in equation (23) no longer apply, and the term only consists of the inverse function of the temperature-radiation correlation of the applied camera:

$$\varepsilon = 1 \text{ is:} \quad \theta_M = M^{-1} \cdot (M_{IR}) \tag{24}$$

Modern thermography systems simplify the evaluation of equation (23) by allowing the input of a radiation temperature $\theta_{U,K}$ (respectively the ambient radiation U or ambient (radiation) temperature θ_U; see Section 2.2.3) and the (assumed) emissivity ε_M of the measuring object by the user. Thus, the only difficulty in measuring the correct temperature remains in the input of the right ambient radiation temperature θ_U and the corresponding emissivity.

Equation (23) is significantly simplified when the temperature of the measuring object θ_M is equal to the ambient radiation temperature θ_U. In this case, the temperature is always displayed correctly regardless of the magnitude of the emissivity. This is because the ambient radiation R_U reflected by the measuring object complements the characteristic

Physical Basics of Thermography

radiation M_M as if the measuring object were a black radiator (since all temperatures are the same). In construction thermography this can be the case when an indoors thermography is conducted and the ambient temperature is almost the same as the temperature of the measuring object.

In most cases, for outdoors thermography, it is sufficient to estimate the ambient radiation on the basis of the exterior temperature. Especially in the case of overcast sky and low clouds, the ambient temperature is relatively close to the exterior temperature.

Determination of the ambient temperature by means of a hemisphere mirror

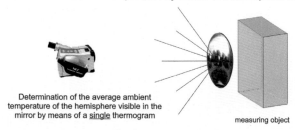

Determination of the average ambient temperature of the hemisphere visible in the mirror by means of a <u>single</u> thermogram

measuring object

Determination of the ambient temperature by means of measuring step-by-step

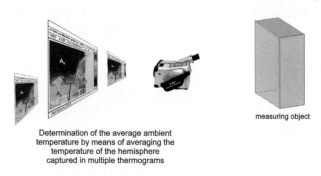

Determination of the average ambient temperature by means of averaging the temperature of the hemisphere captured in multiple thermograms

measuring object

Determination of the ambient temperature by means of a fish-eye objective

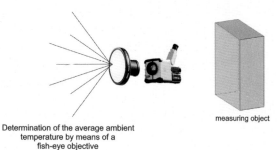

Determination of the average ambient temperature by means of a fish-eye objective

measuring object

Fig. 2-25 Approaches for determinig ambient (radiation) temperature θ_U.

For analyses that require an exact absolute temperature, the radiation temperature of the background must be measured. Here, the ambient temperature of the half-space in front of the object to be analyzed is determined by means of an additional measurement. The determination of the background temperature can be conducted with the aid of a so-called IR fish-eye objective lens (interchangeable lens). A simple, practical approach is to measure the temperature of surrounding areas (such as buildings, sky, trees, and the ground) and calculate the average temperature under consideration of the area percentage.

In another approach, the temperature is determined with the aid of a theoretical hemisphere mirror. The hemisphere mirror is used to project the ambient radiation of the half-space onto the mirror area which is detected by the IR camera. To determine the ambient temperature, the average temperature of the received IR radiation is analyzed by evaluation software (integration of temperatures over the area with possibly different emissivity). In order not to falsify the integration, the area on which the IR camera and the user are detected is excluded from the integration.

A possible modification of the hemisphere mirror principle is the use of creased and resmoothed aluminum foil as a reflector of the background radiation (to use Lambert's cosine law as an ideal diffuse radiator, Johann Heinrich Lambert, 1760), as is explained in [38]. Hereby the creased structure is to reflect the background radiation as diffusely as possible. The background temperature is determined at the foil reflector presuming an emissivity of $\varepsilon = 1$. This background temperature can then be used as input data for the camera for the ambient temperature.

If the main objective of the measurement is not to determine the absolute temperature but to show the temperature differences between the thermographed areas, it is useful to presume a (simplified) emissivity of $\varepsilon = 1$ for the thermograms. In this case, the ambient radiation temperature does not matter for the displayed temperature in the camera (see Section 2.2.3).

2.2.3 Effects of Misjudgment Concerning Emissivity and Ambient Radiation Temperature

On the basis of the general evaluation equation for thermography, wrongly presumed emissivity and radiation temperatures can affect the measuring result. The following examples demonstrate the impacts of such false assumptions.

Base Case

The actual temperature $\theta_M = 0°C$ and the emissivity $\varepsilon = 0.95$ of a surface to be thermographed are known. According to the Stefan-Boltzmann law – see equation (1) – this surface emits about 300 W/m^2.

$$M = 300 = 0.95 \cdot \sigma \cdot T_M^4 \tag{25}$$

Equation (25) can be transposed and solved for the absolute temperature T_M (to the 4th power):

$$T_M^4 = 315 \cdot \sigma^{-1} \tag{26}$$

The exact object temperature T_M can only be determined if the ambient radiation U is given. The data of the ambient radiation then has to be fed to the camera in the form of the corresponding radiation temperature $\theta_{u,\varepsilon=1}$. The amount of radiation emitted by the measuring object that arrives at the thermography camera depends on the actual emissivity of the measuring object and the magnitude of the ambient radiation U, as is shown in Fig. 2-26. The first assumption is that the ambient radiation U = 100 W/m² is less than the specific radiation of the measuring object.

In anticipation of the effect of different ambient radiation values, the distinctive changeover point is the radiation value at the governing temperature of the measured object considering the emissivity of a black body (ε = 1). In the previous example it was

$$M_{\varepsilon=1} = 315 = 1.0 \cdot \sigma \cdot T_M^4 \tag{27}$$

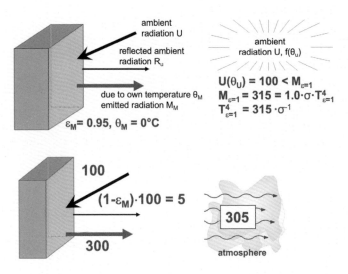

Fig. 2-26 Correlation and values of the radiation balance in order to determine the specific temperature θ_M. Presumption: ambient radiation U(θ_U) = 100 W/m² emissivity ε = 0.95 and specific temperature of measuring object θ_M = 0°C.

On the basis of the correlation between emissivity and reflectance of opaque surfaces – see equation (28) – the values of infrared radiation M_M = 300 W/m² that is emitted due to the specific temperature θ_M = 0°C and ambient radiation R_U can be balanced as depicted in Fig. 2-26.

$$\rho_M = 1 - \varepsilon_M \qquad (28)$$
$$R_U = (1 - 0.95) \cdot 100 = 5 \text{ W/m}^2$$

The radiation M_Σ = 305 W/m², which is comprised of M_M = 300 W/m² from the measured object itself and R_U = 5 W/m² reflected by the ambient radiation.

In order to determine the temperature of the measuring object, data about the presumptive emissivity ε_K of the measuring object and the presumptive ambient radiation temperature $U(\theta_{U,K})$ have to be fed either to the camera or to the evaluation software later on.

Impact of Wrongly Presumed Emissivities ε_K

In the example shown in Fig. 2-27 it is assumed that the emissivity of the measured surface ε_K = 0.7 is considered too low, but through the specification of the camera value a realistic result can be obtained. The ambient radiation U is displayed correctly on the camera. By means of the camera the superimposed shares of the radiation are balanced on the basis of the measured radiation (M_{IR} = 305 W/m²). With the emissivity of the camera settings being presumed too low, the share of the ambient radiation that is reflected by the measuring object is calculated as follows:

$$R_{U,K} = (1 - 0.70) \cdot 100 = 30 \text{ W/m}^2 \qquad (29)$$

The difference of the measured total radiation and the balanced reflected share can be written as: 305 W/m² – 30 W/m² = 275 W/m². The retrograde calculation of the camera based on the Stefan-Boltzmann law provides the following:

$$M = 275 = 0.70 \cdot \sigma \cdot T_K^4 \qquad (30)$$

$$T_K^4 = 392 \cdot \sigma^{-1} > 315 \cdot \sigma^{-1} = T_M^4$$

Comparing the results of the actual object temperature T_M given in equation (30) with the temperature T_K being balanced by the camera software shows that in this case, the temperature T_K that is displayed by the camera is higher than the actual surface temperature T_M.

Physical Basics of Thermography

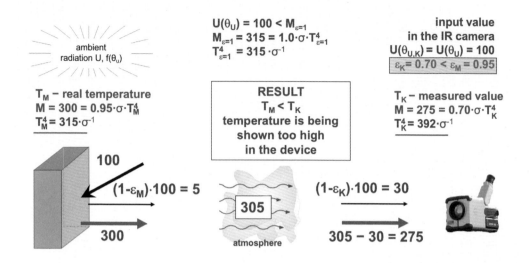

Fig. 2-27 Impact on the results of the radiation balance for determination of the specific temperature θ_M, illustrated for the following case: (1) emissivity ε_k being presumed too low; (2) ambient radiation < max. radiation M, that occurs for the object temperature, assuming ε = 1; and (3) correct input of ambient radiation.

In the case of the assumption that ε_K = 0.99, which is too high compared to the actual emissivity ε_M = 0.95, the temperature displayed on the camera is lower than the actual temperature (see Fig. 2-28).

In the previous cases the ambient radiation was presumed to be lower than the radiation that occurs for the measured object temperature under the assumption of the emissivity of a black body (ε = 1).

In the following, the ambient radiation $U(\theta_U)$ = 400 W/m^2 is presumed to be higher than the emitted radiation $M_{\varepsilon=1}$ = 315 W/m^2, assuming ε = 1. This could be the case when hot radiation sources emit a high ambient radiation toward the measuring object. Under these circumstances an emissivity that is too low can be assumed, which leads to a too low temperature displayed on the camera (see Fig. 2-29). Hence, the resulting effects are opposite to the ones shown in Fig. 2-27.

For high ambient radiation an emissivity being presumed higher than the actual value leads to higher temperatures of the measuring object displayed on the camera (see Fig. 2-30).

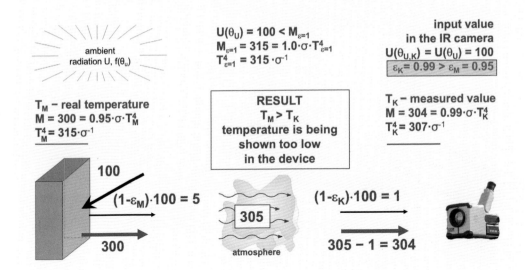

Fig. 2-28 Impact on the results of the radiation balance for determination of the specific temperature θ_M, illustrated for the following case: (1) emissivity ε_K being presumed too high; (2) ambient radiation < max. radiation M, that occurs for the object temperature assuming $\varepsilon = 1$; and (3) correct input of ambient radiation.

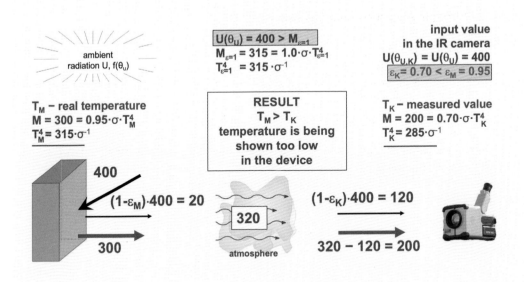

Fig. 2-29 Impact on the results of the radiation balance for determination of the specific temperature θ_M, illustrated for the following case: (1) emissivity ε_K being presumed too low; (2) ambient radiation > max. radiation M, that occurs for the object temperature presuming $\varepsilon = 1$; and (3) correct input of ambient radiation.

Physical Basics of Thermography

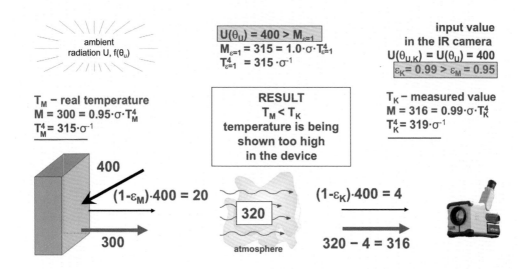

Fig. 2-30 Impact on the results of the radiation balance for determination of the specific temperature θ_M, illustrated for the following case: (1) emissivity ε_K being presumed too high; (2) ambient radiation > max. radiation M, that occurs for the object temperature presuming $\varepsilon = 1$; and (3) correct input of ambient radiation.

Impact of Wrongly Presumed Ambient Radiation $U(\theta_{U,K})$

Since the values displayed on the thermography camera depend on the ambient radiation, the impact of wrongly presumed values for ambient radiation $U(\theta_{U,K})$ is analyzed below. At this point it must be said that the extent to which the presumed ambient radiation differs from the value $M_{\varepsilon=1}$ is irrelevant. For example, if the ambient radiation temperature entered in the camera is too low, the temperature displayed rises (see Fig. 2-31).

If the presumed ambient radiation is too high (see Fig. 2-32) the share of reflected radiation increases as well; the measured value is lower than the actual temperature.

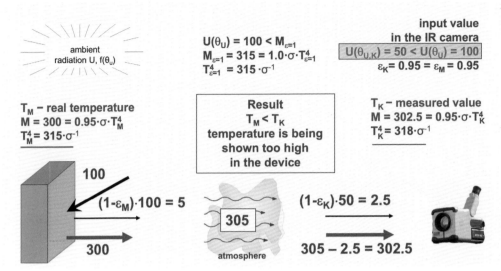

Fig. 2-31 Impact on the results of the radiation balance for determination of the specific temperature θ_M, illustrated for the following case: (1) emissivity ε_K being presumed correctly; (2) ambient radiation < max. radiation M, that occurs for the object temperature presuming $\varepsilon = 1$; and (3) ambient radiation chosen is too small.

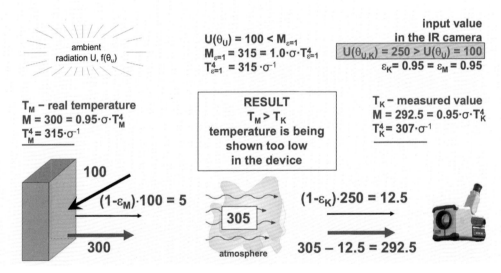

Fig. 2-32 Impact on the results of the radiation balance for determination of the specific temperature θ_M, illustrated for the following case: (1) emissivity ε_K is being presumed correctly; ambient radiation < max. radiation M, that occurs for the object temperature, assuming $\varepsilon = 1$; and (3) the ambient radiation chosen is too high.

Impacts in the Case of Ambient Radiation Temperature $U(\theta_U)$ Being Equal to $M_{\varepsilon=1}$

If the magnitude of the ambient radiation is the same as the magnitude of the radiation that would occur for the actual measuring object temperature and the emissivity of a black body ($\varepsilon = 1$), the measuring result is independent of the input emissivity. Under these circumstances, the surface temperature is measured correctly for every case, independent of the input emissivity (see Figs. 2-33 and 2-34).

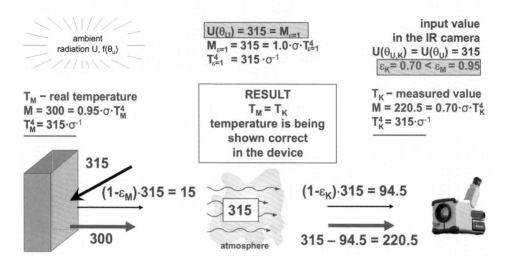

Fig. 2-33 Impact on the results of the radiation balance for determination of the specific temperature θ_M, illustrated for the following case: (1) emissivity ε_K assumption is too low; (2) the ambient radiation = max. radiation M, that occurs for the object temperature presuming ε = 1; and (3) the correct input of ambient radiation.

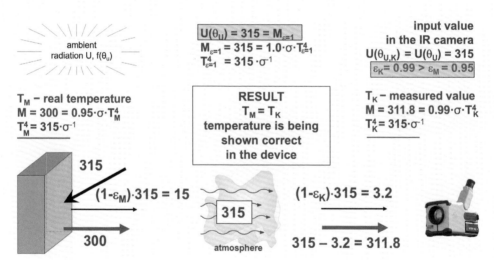

Fig. 2-34 Impact on the results of the radiation balance for determination of the specific temperature θ_M, illustrated for the following case: (1) the emissivity ε_K assumption is too high; (2) the ambient radiation = max. radiation M, that occurs for the object temperature presuming ε = 1; and (3) the correct input of ambient radiation.

3 Overview of Thermography Systems

3.1   Historical Development

In the year 1800 the astronomer William Herschel (born 1738 as Friedrich Wilhelm Herschel in Hannover, Germany) discovered nonvisible heat radiation by scattering sunlight through a prism and proving the existence of heat radiation with the aid of a thermometer. In 1830 the first thermoelements were invented; the so-called radiation thermopile (arrangement of a series of thermoelements) was realized.

In 1880 infrared radiation was proven for the first time by means of changing electrical resistance in dependency of the temperature.

In 1920 quantum detectors were invented which were capable of measuring infrared radiation without first transforming the radiation into temperature. Instead, the striking photons or quantum of energy are directly used for a signal generation with the detector. This invention was then used by the military and applied for infrared guided missiles in 1950.

In 1954 the first imaging camera was introduced. Depending on the receiver system, the exposure time of this camera was between 4 and 20 minutes. In 1964 a scanner camera with a cooled indium antimonide (InSb) receiver was developed in Sweden. This camera only needed a sixteenth of a second to generate an image.

Around 1980, the technology developed by the military could be used for the refinement of quantum receivers. Cooled mercury cadmium telluride (HgCdTe) was used as receiver material and was applicable for spectral ranges of 3–5 μm or 8–14 μm respectively.

In the 1990s high-resolution scanner cameras which were mostly cooled by liquid nitrogen (LN_2) came onto the market. Thanks to the progress in computer technology and related software development, thermography has become more user-friendly.

Since 2000 more and more thermography systems with uncooled thermal receiver arrays, so-called focal plane array (FPA) systems, have come onto the market. The advantage of these systems compared to one-element receivers is the matrix of many individual receivers that function without opto-mechanical scanning. For many applications, these systems provide a reasonable thermal resolution.

Moreover, systems equipped with cooled quantum detectors (HgCdTe or QWIP detectors) are available. By means of these systems, high measuring accuracies (< 10 mK) are possible and fast processes can be captured (refresh rate > 1000 Hz). Nowadays, the trend is toward larger receiver arrays, better temperature resolution, and higher refresh rates. Moreover, more and more thermography systems are equipped with integrated image systems displaying the visible range.

3.2   Camera and Sensor Technology

After the first thermal imaging devices came onto the market between 1960 and 1965, the objective of practical application of infrared imaging in civil engineering was to measure the intensity of emitted infrared radiation and hence to determine the surface temperature of constructions.

Modern thermography systems can be classified by various characteristics. In general (according to DIN EN 16 714-2 [19]), there are two types of camera systems:

* scanning camera units with a single detector or a detector bar
* FPA – cameras with a receiver matrix

Generally, the main detector types (receivers) used can be classified either as *thermal detectors* or as *quantum detectors* (the latter are also known as *photon receivers* or *semiconductor receivers*). Furthermore, thermography cameras can be classified by the spectral range they measure in (according to DIN EN 16714-2 [19]):

* shortwave cameras (SWIR), by definition from 0.8 μm to 2.0 μm
* midwave cameras (MW), by definition from 2.0 μm to 5.0 μm
* longwave cameras (LW), by definition from 8.0 μm to 14.0 μm

Thermal detectors such as microbolometers or pyroelectric detectors have the advantage of functioning at room temperature (approx. 300 K). Incident radiation causes a warming that changes the electrical signal received by the detector (in a bolometer, it changes the resistance; in a pyroelectric detector, it changes the charge).

In order to detect infrared radiation, thermal detectors don't have to be cooled. Some cameras have a built-in temperature stabilizer. The disadvantage of thermal detectors compared to quantum detectors is the low detection capability of the radiation (specific detectivity). Due to the thermal mass of the detector materials, it takes time to detect the radiation, which makes measuring sudden temperature changes (which is important for industrial use and research) difficult. The characteristic value of the thermal resolution of the camera – that is, the smallest distinguishable temperature difference (NETD, see Section 3.3) – is about 0.1 Kelvin or better for these kinds of detectors.

Quantum detectors use the inner photoelectric effect to detect infrared radiation, which means they function as photon counters. When radiation hits the detector, the resistance changes immediately. Thus, quantum detectors provide instant measurements. Moreover, quantum detectors have a higher specific detectivity (sensitivity) compared to thermal detectors and generally exhibit a low-noise image. However, these properties can only be achieved by cooling to a very low operating temperature (77 K = −196°C). The characteristic value of the thermal resolution (NETD, see Section 3.3) of these detector types reaches to about 0.01 K (and better).

Currently, the thermography systems available on the market are equipped with a fixed lens with focusing device. The optical field of view can usually be adapted to the respective requirements by means of attachable additional lenses of fixed focal length. This can be necessary, for example, if long distances have to be bridged (telephoto lens) or if the field of view has to be extended in case of space problems or if objects with a high depth of focus have to be focused particularly closely (wide-angle lens).

3.2.1 Scanner Thermography Systems

Scanner cameras are equipped with a single-element detector. In order to create an image, an opto-mechanical deflection system (scanner) has to form a grid of the considered object (scan the object in a grid-like form). There are several possible setups of such a scanner system. In the literature, scanner setups are described by plane mirrors (oscillating mirrors), mirror polygons, or polygon prisms [36]. Regardless of the chosen setup, they all scan the object vertically and horizontally to create an image line by line.

The focused radiation is sent to an infrared-sensitive single-detector or detector band (detectors arranged linearly). There, the signals are detected, amplified, and eventually assembled into an image by means of software.

The advantage of scanner cameras is the high thermal resolution. As for single-element receivers, every single picture element is transferred into a temperature by the same detector. This leads to a very homogeneous thermography image. However, due to the opto-mechanical scanning process, the imaging speed of these cameras is slow (about 1 second per image).

Usually, the detectors used for scanner cameras are quantum detectors, such as mercury cadmium telluride (HgCdTe = CMT).

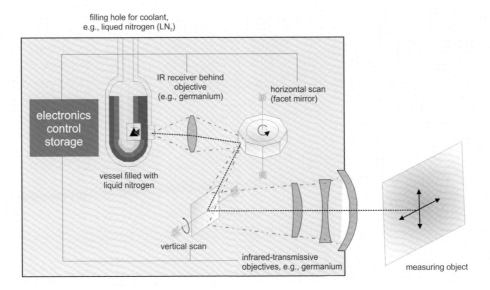

Fig. 3-1 Schematic illustration of the imaging process of a nitrogen-cooled scanner camera.

The choice of an appropriate detector depends on the radiation maximum in a certain wavelength range. For good detectivity of the receiver and to minimize background emission (self-emissions) in the considered wavelength range, it is necessary to cool the detector. The following three cooling principles are commonly used in construction thermography:

- open cooling with liquid gas (mostly nitrogen; in the lab: helium) in an upright Dewar flask;
- cooling by means of a Stirling-type cooler;
- thermoelectric cooling with Peltier elements.

Liquid Gas Cooling

Liquid nitrogen is commonly used in construction thermography. Liquefied gases are filled into a Dewar flask.[3] The detector that is to be cooled sits on the bottom of the Dewar flask, positioned behind an infrared-pervious screen so radiation can get to the detector. The evaporation heat of the liquid gas keeps the receiver constantly at the boiling point ($-196°C$ = 77 K for nitrogen). One disadvantage of this method is the handling of the liquid gas (transportation and storage) and the compulsory upright position of the Dewar flask, which restricts the view. Scanning an object from a slanted angle of view, as is often necessary in construction thermography, is not possible. Liquid gas cooling allows a usage time of about four to ten hours for most thermography cameras.

Fig. 3-2 Cooling with liquid nitrogen: A portable tank is filled with liquid nitrogen from a storage tank (*left*) and used to fill a thermography camera (*right*).

---

3 Named after Sir James Dewar, who invented the double-walled, thermally insulating vessels (vacuum jacket and silvering).

Cooling by Means of a Stirling Cooler

The Stirling cooler is a mechanically driven cooling machine that generates cold by thermo-dynamic cyclic processes. The invention of the Stirling cooler was primarily advanced by the military and the aerospace industry. In order to reduce vibration load, Stirling coolers are equipped with a linear drive. Furthermore, the operation chambers are split so that the cold is generated in only one of the two cylinders ("cold finger"). The working gas is transported between the two cylinders by a connection pipe. Fig. 3-3 exhibits such a split Stirling cooler. Stirling coolers generate a refrigerating capacity of 1–2 Watts out of a consumption power of 40–90 Watts. When bringing a thermography camera equipped with a Sterling cooler into service, it takes about ten to twenty minutes to generate the operating temperature (77 K). Due to the mechanical wear of Stirling machines, the operating time of such a cooler is about 50000 to 60000 hours.

Fig. 3-3 Stirling cooler: linear-driven, split model (Image: Fa. AIM GmbH, Heilbronn 2003), 1.5 Watts refrigerating capacity (left: cold tip = cold finger, Ø 13.7 mm).

Cooling with Peltier Elements

The principle of thermoelectric cooling is based on the Peltier effect.[4] When certain metals are brought together conducting direct current, one of the contact points cools down while the other one heats up. Peltier elements can be arranged in multiple stages in order to amplify the cooling process. Peltier elements are not capable of generating temperatures as low as the methods discussed above, yet they are very reliable and durable.

Fig. 3-4 Sectional view of an infrared receiver, illustration of a multistage Peltier cooling element. Located on top is the infrared detector; the case is covered with an infrared-pervious screen in the visual range of the detector (Photo: Fa. AIM GmbH, Heilbronn).

---

4 The thermoelectric effect is named after its discoverer, J. C. Peltier, published in 1834.

3.2.2 Focal-Plane-Array Thermography Systems

In the last few years, Focal-Plane-Array (FPA) cameras have increasingly been implemented for thermography systems. Here, the incident infrared radiation is focused on a detector matrix with many individual receiver sensors. The line-by-line scanning of the image and the corresponding opto-mechanical technology do not apply for this camera type.

The advantage of the FPA camera is the rapid image capture. Due to the many single receivers of the matrix and the resultant characteristic transducer lines, irregularities of the different measuring points may occur. For characterization of the detector matrix, a filling factor can be used. The filling factor describes the ratio of the area that is sensitive to radiation to the total area of the FPA. FPAs with a high filling factor display a smoother, more homogeneous image – yet they can be prone to cross-fading (mutual influence of neighboring detectors). The negative effects described above are compensated by built-in nonuniformity corrections (NUC) and the thermal stabilization of the detector.

Depending on the operation purpose and the required accuracy, FPA cameras are offered with different detector principles (thermal detector, quantum detector) and as cooled or uncooled cameras respectively. Under certain circumstances, FPA detectors underlie the legislation on export control of dual-use goods and technologies so that restrictions on personalized use, for instance, are given. Due to export regulations information about the absolute accuracy of FPA detectors is also provided.

The rapid development of uncooled thermal FPAs since about the beginning of the 2000s/ 2010s has led to an improvement in the geometrical resolution up to 1280 × 1024 pixels and a thermal resolution, that is, the smallest distinguishable temperature difference of about 0.02 K (20 mK) or better for building thermography. Moreover, the application of MEMS (Micro Electronic Mechanical Systems) has also allowed other systems that can generate a higher geometrical resolution to become market-ready.

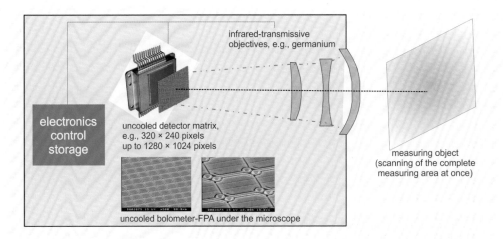

Fig. 3-5 Imaging of an FPA camera.

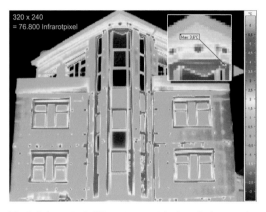
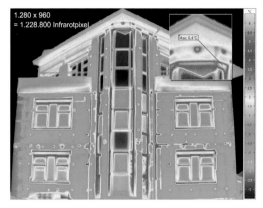

Fig. 3-6 Impact of different geometrical resolutions on thermograms of a building: The very high resolution in the thermogram on the right is achieved by means of an additional technology (resolution enhancement, MicroScan). Source: With the friendly assistance of InfraTec GmbH (www.InfraTec.net).

The (repeat) measurement accuracy is usually specified by the manufacturer, with the higher value of ±1°C or ±1% of the rank. Basically, with increasing geometric and decreasing thermal resolution, the thermogram becomes clearer and more detailed. Details are also easier to recognize in more distant objects (see Figs. 3-6 to 3-8).

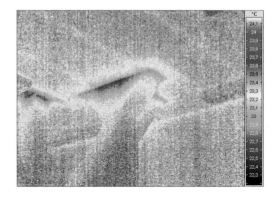
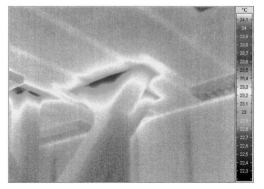

Fig. 3-7 Comparison of a construction being thermographed with different thermal resolutions: (left) 80 mK; (right) 30 mK. Source: With the friendly assistance of InfraTec GmbH (www.InfraTec.net).

Thermal image with a resolution of 140 x 140 pixel and a thermal sensitivity of <100 mK

Thermal image with a resolution of 200 x 150 pixel and a thermal sensitivity of <70 mK

Thermal image with a resolution of 320 x 240 pixel and a thermal sensitivity of <50 mK

Thermal image with a resolution of 640 x 480 pixel and a thermal sensitivity of <45 mK. The higher number of pixels results in a very clear picture in which small details are visible.

Fig. 3-8 Thermograms of a residential building to illustrate the impact of different thermal and geometrical resolutions. Source: With the friendly assistance of FLIR Systems GmbH (www.flir.com).

3.3   Criterion for Thermal Resolution

When choosing a thermography camera, the choice of the temperature measurement range is very important. The temperature measurement range describes the lowest temperature to the highest temperature measurable by the IR camera. For reasons of comparability, the temperature measurement range should be based on the measurement of a black body. A suitable temperature measurement range of IR cameras used for construction thermography is between −30°C and +100°C. If for any applications special requirements have been set, the temperature measurement range can be extended or adjusted by professional suppliers of IR systems.

IR cameras can be specified based on their thermal resolution. The thermal solution indicates the smallest temperature difference that can be distinguished by the IR camera. Temperature differences or radiation differences below the thermal resolution limit cannot be displayed by the IR system. Thus, a better thermal resolution (in practice) leads to a more homogenous, clear, and distinct image.

The international abbreviation for thermal resolution at an object temperature of 30°C (resolution varies with object temperature!) and a specific refresh rate without averaging is NETD (Noise Equivalent Temperature Difference [K]).

Overview of Thermography Systems

For current (as of 2021) thermography systems a common NETD is about 0.015 K (for cooled quantum detectors) to < 0.02–0.1 K (for uncooled thermal detectors).

3.4   Optical Elements: Lens Systems

Thermography systems include optical elements such as lenses, mirrors, prisms, etc., which create a minimized image of the emitted infrared radiation on the receiver system. As explained in the physical basics of thermography in Chapter 2, common optical materials made of glass are impervious to longwave heat radiation. Therefore, lenses and optical protection screens of thermography systems are made of materials that are pervious to infrared radiation, such as silicon (Si) or germanium (Ge). In order to achieve higher transmittances, additional antireflection coatings for the lenses can be used. Mirrors aiming to completely reflect the incoming radiation are made of polished metals or made with gold sputtering.

For the characterization of the optical system of an IR camera, the following parameters are given:

FOV (Field of View [Degree])

FOV describes the image section captured by an IR system (camera and objective) and depends on the camera type (scanner principle or FPA principle) and the objectives used (focal width). The FOV of IR systems used for construction thermography normally ranges

Fig. 3-9 Comparison measurement with thermography cameras: (*left*) scanning single-element camera equipped with a Stirling cooler; (*right*) uncooled micro-bolometer camera, with 320 × 240 pixels resolution (state-of-the-art in 2001).

from 20° to 30° (horizontally) and 15° to 20° (vertically). The field of view shrinks with decreasing FOV values and enlarges with increasing FOV values.

The thermography systems that are currently on the market are generally equipped with a fixed objective with focusing mechanism. The optical field of view can then be adjusted by means of mountable additional objectives with fixed focal width. This could be necessary if greater distances have to be spanned (use of a telephoto lens) or if the camera has to be placed so close to the object (for instance, because of restricted space) that the field of view has to be extended with high depth of focus (wide-angle lens). Since interchangeable lenses are rather expensive, it should be mentioned that there is a software offered by one supplier, which assembles the single images into one overall image. Thereby the radiometric data of the subimages is completely inherited, so that the assembled overall image can still be edited later on.

IFOV (Instantaneous Field of View [mrad])

The IFOV or "detector footprint" is the field of view of one single detector element. It specifies how big a single dot (in [mm]) is in the generated image when the measured object is at a distance of 1 m. Each of these dots is assigned to one detector. The geometrical size of the smallest measuring dot is then calculated by multiplying the value of the IFOV [mrad] with the object distance in [m] and a correction factor considering the lenses used.

The exact values of the geometrical resolution depending on the camera type and lens can be taken from the technical instructions provided by the manufacturer of infrared systems. Camera systems that are commonly used for building or construction thermography should have an IFOV of < 1.5 mrad [30] or task-related [40] respectively. Scanner camera systems allow for the generation of electro-optical zoom, depending on the scanner. With electro-optical zoom the image can be enlarged while the measuring dot size is retained.

Moreover, the geometrical resolution of scanner camera systems or other cameras in interaction with the objective is indicated by means of the Split Response Function (SRF) (see [19]). Fig. 3-10 illustrates the field of view (FOV) and the instantaneous field of view (IFOV) as they are provided in the manufacturer's data of an FPA camera system.

The information about the IFOV or the size of the measuring dot is of special interest when small "warmer" details have to be determined exactly over a long measuring distance by means of thermography. In case the measuring dot is greater than the warmer detail, the detector averages the residual area of the measuring dot with the radiation of the cooler area. For more exact measurement of the detail, the measuring dot must not be greater than the detail itself. In fact, for better results the measuring dot should be much smaller than the detail. In Chapters 7.5.2 and 8.2.1, an example is used to visualize the impact of the spacing effect.

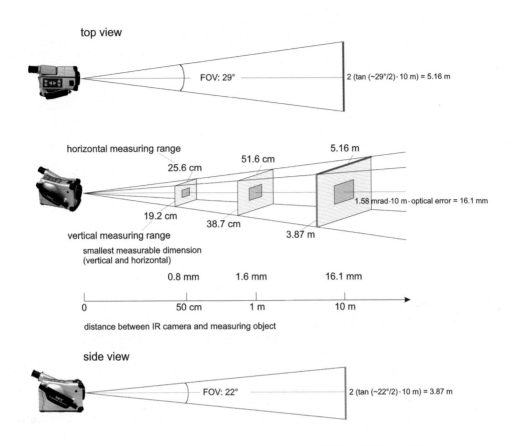

Fig. 3-10 Illustration of the geometrical specification of IR camera systems (FOV and IFOV), see [31].

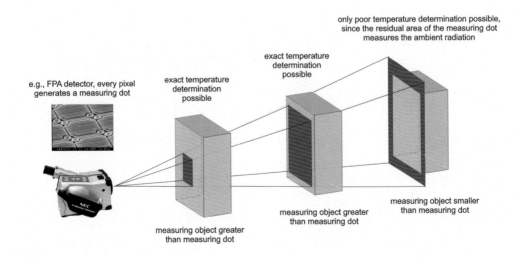

Fig. 3-11 Explanation of measuring dot size.

3.5   Illustration of Thermograms, Storage, and Processing

Thermograms are fully radiometric, that is, all information of the detected radiation in every pixel is recorded. In order to provide a clear visualization and evaluation of the recordings, different colors are assigned to the calculated temperatures, which generates a so-called off-color diagram.

The colors of the thermographs are chosen in such a manner that details are depicted optimally. However, this also means that in different thermograms the same colors don't necessarily indicate the same temperature. As a general convention, black, blue, and green colors are used to indicate low temperatures, whereas yellow, orange, and red colors are used for higher temperatures. Temperatures that are higher or lower than the defined temperature range can be indicated by white or black areas.

Depending on the purpose of use, manufacturer-specific charts of color range are offered. Since the maximum and minimum of the temperature range are freely selectable, it is possible to display differences of the surface temperature of < 1 K in such a way that even small temperature differences seem like significant heat loss. Thus, the color temperature range and the boundary conditions of the measurement must be considered when evaluating a thermogram. When taking a series of thermograms, the color temperature range should be chosen uniformly for all thermograms to allow for a good comparability.

Fig. 3-13 shows a selection of different temperature scales available for the software used to generate the thermograms in this book.

Since the temperatures of each pixel in a thermogram are saved individually, the image can also be altered afterward.

So far, the choice of the color scale is not regulated. However, the experience of the last few years has shown that the color effect is often evaluated on an emotional basis, so that

Fig. 3-12 The exact same thermogram with different temperature scales can lead to a different subjective impression.

Alle Paletten

	delta
	he1
	nec_brown
	nec_ironbow
	nec_spread
	nec_standard
	orig
	pwi_fire
	pwi_spreaddelta
	pwi_spreadhigh
	pwi_spreadlow
	pwi_spreadrgb
	pwi_spreadsuper
	pwi_standard
	rgb
	schema1

Fig. 3-13 Temperature scales available (for the software PicWin-IRIS).

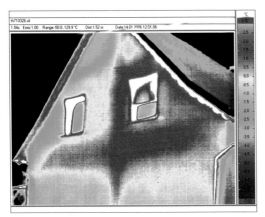 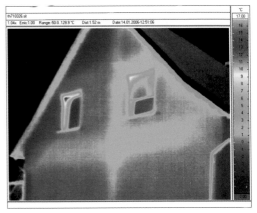

Fig. 3-14 The exact same thermogram with different temperature scales can lead to a different subjective impression.

especially unexperienced users associate the color red with a negative result, even though there is no substantive basis for that interpretation. In addition, as more service providers have realized that a standardized neutral illustration would simplify the comparability of thermograms, such standardization is under discussion.

In order to realize such standardization, a government-funded research project was founded in Switzerland to develop an evaluation tool (QualiThermo). With the aid of this tool, reference thermograms can be generated that are supposed to allow for comparison of thermograms that were captured under different meteorological boundary conditions (see [5]).

Furthermore it is suggested to apply the following basic rules to obtain illustration neutrality [23]:

- The scale should span about $0.7 \cdot$ temperature difference ($\theta_{interior} - \theta_{exterior}$).
- All thermograms should use the same color range; double seizure of colors is to be avoided.
- For outdoors thermography, the exterior temperature should be at about 20% of the temperature scale.
- For indoors thermography the interior temperature should be at about 70% of the temperature scale.

Fig. 3-15 shows a thermogram scaled according to the suggestions above. It remains to be seen if the suggestions for a standardized illustration of thermograms will be adopted in practice.

Date	Temperature boundary conditions		Additional information
	Exterior temperature	Interior temperature	
14.01.2006 about noon	approx. −4°C	approx. +22°C	Cold weather for about 8 days Application of the suggestions for illustration neutrality according to [23], Temperature difference: 26 K Scale: 0.7 · 26 K = 18.2 K (−7.6°C to +10.6°C) Exterior temperature at about 20% of the temperature scale

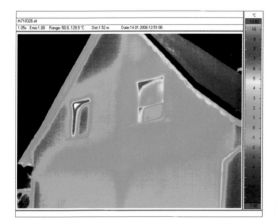

Fig. 3-15 Thermogram scaled under application of suggestions for illustration neutrality of thermograms according to [23].

If damage-free buildings with a high thermal insulation standard (such as the low-energy standard) are thermographed and processed in accordance with the design rules for presentation neutrality, the thermograms show only very weakly recognizable thermal structures. The somewhat warmer frames of the windows are often recognizable as typical features. The surface temperatures correspond approximately to the prevailing ambient temperature.

Overview of Thermography Systems

Date	Temperature boundary conditions		Additional information
	Exterior temperature	Interior temperature	
28.01.2007 about noon	approx. +1°C	approx. +20°C	covered sky, application of the design rules for neutral display of thermograms

Fig. 3-16 Thermogram scaled under application of suggestions for illustration neutrality of thermograms according to [23]. Outdoor thermography of a damage-free, low-energy house with modern thermal insulation standard. Usually only warmer window frames are thermally visible. As a rule, the wall area shows no structure.

4 Regulations, Guidelines, and Standards

Various organizations, associations, and standardization bodies have produced several documents on the subject of building thermography. These documents are presented below.

In 1983 the international standard ISO 6781 [27] was published in order to evaluate the thermomechanical behavior of buildings and construction parts; it deals with the verification of thermal bridges by means of infrared technology. This original standard was then adapted to the ongoing development and modified according to the European requirements. The current version of this standard is the DIN EN 13187, version of May 1999 [14]. This standard mainly focuses on the basic principle of thermographic examination of building parts and the conduction and analysis of thermograms. This standard, edited by the construction committee of the DIN, also provides specific guidance that has to be followed when thermographing building envelopes.

Another standard on the subject of thermography is the DIN EN 16 714 Thermographic testing with Parts 1 to 3 [18][19][20], which was published in November 2016 as a replacement for DIN 54 190. This standard focuses on the general basics of thermographic examination (that is, procedure, techniques, personnel requirements, etc.) and on equipment technology (such as the technical details of IR cameras and technical requirements concerning the camera systems) and explains important terms concerning thermography. Due to developments in the field of nondestructive testing, the DIN EN 17119, version of October 2018 [21] was introduced. It discusses the procedure of active thermography, possible types of stimulation, analysis techniques, and the conduction of the testing.

For completeness, it should be mentioned that general information on the radiation physics of infrared thermography is contained in the corresponding parts of the standard DIN 5031 [13] and in great detail in VDI/VDE 3511, Sheet 4 – Technical temperature measurements, radiation thermometry [41].

Furthermore, reference is also made to the standards published by the ASTM (American Society for Testing and Materials), such as ASTM C 1046 [1], ASTM C 1060a [2], and ASTM D 4788 [3]. In these publications references are given for special practical application areas of thermography.

Further notes and information sheets on the application of thermography in civil engineering have been published in research reports (such as [30]) or drawn up by professional associations (for example, VATh Verband für angewandte Thermografie e.V. [40]; Merkblätter Deutsche Gesellschaft für Zerstörungsfreie Prüfung DGZfP [10][11]; and Wissenschaftlich-Technische Arbeitsgemeinschaft für Bauwerkserhaltung und Denkmalpflege, WTA Merkblatt: Infrared thermography for buildings [48]).

In addition, professional associations, external institutions, and certification societies offer courses in which experienced thermographers can prove their qualification and experience through an accredited certification company. Here, too, there are standards such as DIN EN ISO 9712 – Qualification and certification of personnel involved in nondestructive testing [16] and, especially for infrared thermography, DIN EN ISO 6781-3 [17].

5 Accuracy of Temperatures Indicated in Thermograms

For given/known boundary conditions, nondestructive and noncontact measuring by means of thermography is considered to be very accurate. However, since in most cases thermography is carried out under transient climate boundary conditions, the detected values have to be considered as instantaneous values. According to the regulations of DIN EN 13187 [14], thermograms that are taken in situ have to be compared to the expected temperature distribution across the surface. In order to determine the expected temperature distribution, it is important to know the thermomechanical properties of the building envelope as well as the relevant environmental conditions at the time the thermogram was taken.

Many times, when analyzing thermograms, it has to be questioned how changing the time of recording would affect the measuring results. This is especially important since the temperature fields arising in the wall strongly depend on the mass and the thermal capacity of the construction material. Furthermore, what amount of deviation is to be expected when emissivity and ambient radiations are not sufficiently well known or wrongly assumed must be examined.

5.1   Effects of Climate Boundary Conditions and Age of the Building

Effects on the Size of Measuring Differences at Thermal Bridges

Within the framework of a diploma thesis written at the Leibniz Universität Hannover, the impact of the climate and constructional boundary conditions have been extensively studied by means of an exemplary construction [24]. For this purpose, a theoretical study was carried out, performing variation calculations on the basis of transient calculations of a 3-D construction model.

Classification of the Weather with Regard to Thermography Boundary Conditions

The values of the outdoors climate were based on the data of the test reference year published in 2011, average values and climate region 3 [8]. The weather data were then summarized by means of the so-called fictive air temperature approach, published in [32].

The fictive exterior temperature can be determined as follows:

$$\theta_{fic,out} = \theta_A + \frac{\alpha_{BT} \cdot G(\gamma_F,\alpha_F,N) + \varepsilon_{BT} \cdot (A(\gamma_F,\alpha_F,N) + A_R(\gamma_F,\alpha_F,N) + E_K) - \varepsilon_{BT} \cdot \sigma \cdot T_A^4}{h_r + h_c} \quad (31)$$

with:

$\theta_{fic,out}$	fictive exterior temperature [°C]
θ_A	air temperature [°C] (measured 2 m above the ground)
α_{BT}	shortwave absorption coefficient of the construction surface [-]
$G(\gamma_F,\alpha_F,N)$	global radiation intensity of a tilted and randomly oriented surface with a total degree of coverage N [W/m²]
ε_{BT}	longwave emissivity of the construction surface [-]

$A(\gamma_F, \alpha_F, N)$ radiation intensity of the atmospheric back radiation on a tilted and randomly oriented surface [W/m²]

$A_R(\gamma_F, \alpha_F, N)$ reflected radiation intensity of the back radiation on a tilted and randomly oriented surface [W/m²]

E_K radiant flux density of the longwave emission of the terrestrial surrounding [W/m²]

σ Stefan-Boltzmann constant [$5.67 \cdot 10^{-8}$ W/(m²K⁴)]

h_r heat transfer coefficient due to radiation [W/(m²K)], here: nonvarying approach with: h_r=5 W/(m²K)

h_c heat transfer coefficient due to convection, approach according to DIN EN ISO 6946 [W/(m²K)] [22]

V_{Wind} Wind velocity [m/s]

The result of the approach of fictive temperatures is a single value that balances the effects of shortwave solar radiation, clouds, and longwave atmospheric back radiation. The fictive exterior temperature represents a simplified exterior temperature that would cause the same effects on the considered outdoor surface (warming or cooling), as the single parts of the different radiations would do, if calculated separately.

The results of these calculations are temperature sequences that can vary significantly, especially when the sky is clear. In this case, the effects of solar radiation during the day and a possible negative radiation at night can lead to remarkable deviations between the

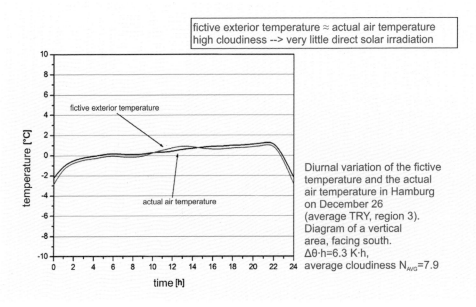

fictive exterior temperature ≈ actual air temperature
high cloudiness --> very little direct solar irradiation

Diurnal variation of the fictive temperature and the actual air temperature in Hamburg on December 26 (average TRY, region 3). Diagram of a vertical area, facing south. $\Delta\theta \cdot h$=6.3 K·h, average cloudiness N_{AVG}=7.9

Fig. 5-1 Sequence of the fictive exterior temperature to the actual exterior temperature, example of a heavily overcast day.

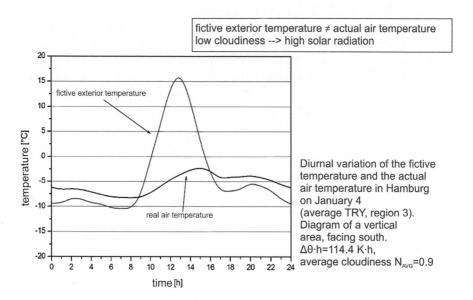

Fig. 5-2 Sequence of the fictive exterior temperature to the actual exterior temperature, example of a slightly cloudy day.

calculated fictive air temperature and the measured exterior temperature. By balancing the magnitude and duration of the deviation, numerical values for a classification of good, suitable, and unsuitable boundary conditions for thermography can be derived.

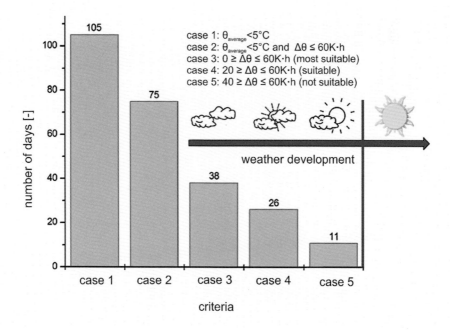

Fig. 5-3 Distribution of suitable days for thermography measurement and corresponding weather development.

Effects of Climate Boundary Conditions and Age of the Building

On the basis of the approaches shown above and the requirement of thermograms to be taken at average exterior temperatures of $\theta_e < +5°C$ to assure a sufficient temperature gradient, the climates have been categorized according to classification criteria. The selected total deviation $\Delta\theta \cdot h$ of 60 Kh equals an average deviation of the fictive exterior temperature from the real air temperature of 2.5 Kelvin per hour.

The results of the calculations based on the TRY (test reference year), zone 3 are shown in Fig. 5-3. It can be seen that for the considered region called "Nordwestdeutsches Tiefland" (lowlands in northwestern Germany) and the chosen classification, there are about 75 days that provide decent preconditions to carry out thermographic measurements.

Practice has shown that stable (bad) weather with constant temperatures around 0°C and overcast sky are very good conditions for thermography measuring. Analyzing Fig. 5-3, such weather conditions are given on 38 single days.

For very cold weather, the capacity of air to hold back humidity and the related low degree of cloudiness create the ideal winter weather. However, for thermography measuring, this weather condition is unsuitable because of the distracting effects of solar radiation. From the available climates of each classification, the dates listed in Tab. 5-1 have been selected to perform the computational analysis.

Tab. 5-1 Selected dates for the simulation of transient boundary conditions.

Day	Data	Classification
December 26	Total deviation $\Delta\theta h$ = 6.3 K·h Average air temperature θ_{av} = 0.3°C Average degree of cloudiness N_{avg} = 7.8	very well suited
January 15	Total deviation $\Delta\theta h$ = 30.8 K·h Average air temperature θ_{av} = 0.1°C Average degree of cloudiness N_{avg} = 6.5	well suited
February 1	Total deviation $\Delta\theta h$ = 57.0 K·h Average air temperature θ_{av} = −5.1°C Average degree of cloudiness N_{avg} = 4.0	sufficient
January 4	Total deviation $\Delta\theta h$ = 114.4 K·h Average air temperature θ_{av} = −5.6°C Average degree of cloudiness N_{avg} = 0.9	unsuitable

Construction

Studies within the framework of a diploma thesis were carried out on the basis of a construction with a wall corner pointing inward and a thermally unseparated ceiling construction (this is further described in Chapter 7.1.1). The chosen construction is a good example of 1960s state-of-the-art building techniques. The load-bearing masonry walls are made of sand-lime bricks (λ = 0.70 W/(mK)), plastered on the interior and exterior. Additionally, clinker face bricks are incorporated in the balcony area. The heat transition coefficient of the exterior wall is about U = 1.5 W/(m²K).

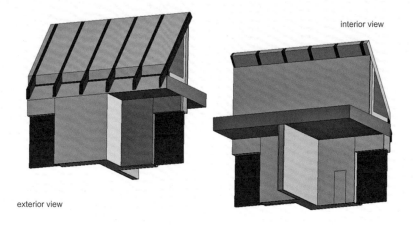

interior view

exterior view

Fig. 5-4 Interior and exterior views of the FE-model of the construction (FE-software: Ansys-Workbench).

In order to examine the influence of construction types and insulation standards on the calculation result, several variations have been tested on the existing construction. For the simulation of a construction typical for the mid-1990s, the existing masonry has been replaced by a brick material with better thermal insulation properties (λ = 0.39 W/(mK)). As a result, the heat transition coefficient of this variation is about U = 0.88 W/(m^{2}K). Since the construction itself isn't subject to any alteration, the continuous concrete ceiling remains (a negative influence) the same as in the original version of the model.

Within the framework of a further modification, the original model was changed in such a way as would be the case for a thermomechanical renovated old building. Thus, the addi-

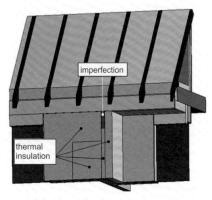

imperfection

thermal insulation

Geometry of the imperfections
- height: 500mm
- width: 5-10-25-50-75mm
- depth: thickness of the
          insulation panels: 160mm

visualization of the building corner with an overhanging concrete slab, used to analyze the impact of imperfections within the insulation layer on the temperature distribution. For reasons of visualization, some parts of the plaster (gray) are being suppressed.

Fig. 5-5 Exterior view of the FE-model with position of the assumed imperfection within the plane of the thermal insulation composite system (software system: Ansys-Workbench).

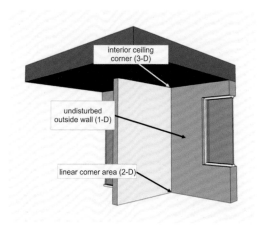

Fig. 5-6 Interior view indicating the characteristic reference points.

tional thermal insulation composite system (d = 16 cm, λ = 0.040 W/(mK)) leads to a heat transition coefficient of U = 0.26 W/(m²K). Besides the exterior wall, the underside of the overhanging concrete ceiling is also improved by means of thermal insulation. In the area of the eaves/rain gutter, the insulation of the exterior walls goes up to the insulation between the rafters, so that an optimized insulation plane is created for the whole building. Moreover; the influence of the structural thermal bridge in the form of the overhanging concrete ceiling is minimized. Within the framework of renovation measures, imperfections in the insulation plane due to voids or mortar clumps cannot be excluded. Thus, the influence of such imperfections on the interior surface temperature has been calculated (see Fig. 5-5).

The influence of the imperfection was then examined on the interior of the building at the characteristic points of the inner corner of the ceiling (3-D), the linear corner area (2-D), and the undisturbed outside wall (1-D).

5.1.1   Effects of Transient Influences: "Snap-Through Time" of Temperature Changes

Due to transient temperature behavior and the heat storage capacity of the construction, real building constructions exhibit buffered temperature behaviors, especially on interior surfaces. Within the framework of the calculations, it was first examined how long it takes for the interior corner of the ceiling to show any alteration due to the altered exterior temperature. For the construction year categories of 1965, 1995, and 2005, under identical climate boundary conditions, the calculations result in so-called thermal snap-through times of 15–18 hours. In this connection, it also can be seen that the absolute temperatures and the differences between minimum and maximum surface temperatures vary significantly depending on the constructional implementation of thermal insulation (see Fig. 5-7).

As a result of the calculations of the resounding times, the climate conditions of the prior twenty-four hours should be considered when evaluating thermograms. Conversely, if there are any changes in the weather, a temporally delayed snap-through of the extreme values on the interior surface must be expected.

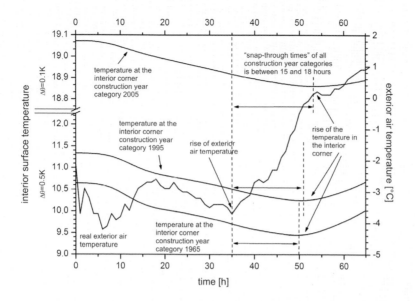

Fig. 5-7 Delayed temperature reaction of the interior corner of the ceiling to the rise of the exterior temperature. Very well-insulated constructions exhibit higher interior surface temperatures and lower thermal fluctuation throughout the day.

5.1.2 Calculation of Daily Fluctuation of Interior Surface Temperatures

To determine the climatic and structural impacts, the temperature sequences that exhibit the highest variation were calculated at the characteristic evaluation points over a period of three days. The outcome on the third day was then used for further evaluation (see Fig. 5-8). Two different cases were distinguished:

- Real case: The climatic boundary conditions two days before the thermography measurement were taken from the test reference year. The thermography measurement was carried out on the third day, that is, the reference day (see Fig. 5-8). The classifications of the first two days are not necessarily the same as those of the third day. The inclusion of the real climates of the test reference year (TRY) two days before the thermography measurement leads to transient effects due to the different classifications of the preceding days.
- Ideal case: The climatic boundary conditions two days before the thermography measurement are presumed to be the same as on the day of the measurement (third day). This approach represents the limit value of the selected classification conditions for the following calculations.

One example of a temperature sequence showing the maximum temperature fluctuation the day of the thermography measurement for the real case and the ideal case on an unsuitable day and the construction year 1995 is shown in Fig. 5-8.

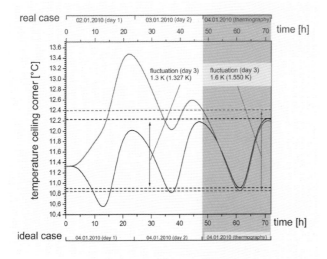

Fig. 5-8 Evaluation principle for determining the maximum daytime-dependent temperature variation for a thermography measurement. Here: construction 1995 on an unsuitable day, according to the classifications in **Section 5.1** and **Tab. 5-1**.

The variations of the interior surface temperatures at the inner corner of the ceiling, according to the classifications, the construction year categories, and the variation of the climate of the preceding days (real case / ideal case) are summarized in Fig. 5-9.

The results of the variation range show that thermograms of old buildings may exhibit absolute temperature variations of about 1 to 2 Kelvin solely because the images were taken at different times of the day. Only for very well insulated buildings (construction year category 2005) is the detectable daytime-dependent fluctuation marginal within the range of the thermography system's measurement inaccuracy. Therefore, the meticulous indication of the measured instantaneous temperatures in thermograms should be reconsidered. Moreover, evaluations based on absolute temperatures have to be calculated considering the ranges of variation.

For analyzing thermograms with regard to thermal bridge effects, the differences of the lowest interior surface temperatures of the inner corner of the ceiling (3-D effect), the temperature along the building edges (2-D effect), and the undisturbed construction (1-D effect) are compared (see Chapter 6.2.4).

The average differences of the considered cases comparing the 3-D corner and the 1-D wall with regard to construction year, climatic boundary conditions, and the variation of the climate of the preceding days (real case / ideal case) are shown in Fig. 5-10. The differences between the temperatures of the 3-D corner and the linear building edge (2-D) are shown in Fig. 5-11.

The calculations show that the temperature differences between the different thermal bridges (3-D compared to 1-D, and 3-D compared to 2-D) vary due to the different construction categories and the associated implementation of the constructions. As a general tendency, the temperature difference of well-insulated constructions is lower

Accuracy of Temperatures Indicated in Thermograms

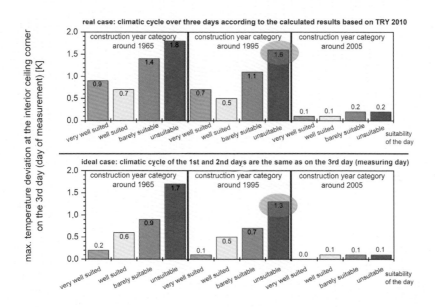

Fig. 5-9 Maximum variation of the temperatures at the inner corner of the ceiling throughout the day, answering the question: "Does the time of day matter in regard to thermography measurements?" The reading values shown in Fig. 5-8 are encircled in the diagram.

than the one of poorly insulated constructions. Tab. 5-2 lists the differences resulting from the calculations.

Tab. 5-2 Differences between the surface temperatures at the characteristic evaluation points (see Fig. 5-6).

construction year category	temperature difference between 3-D and 1-D	temperature difference between 3-D and 2-D
1965	6 to 7 Kelvin	3 to 4 Kelvin
1995	7 to 8 Kelvin	5 to 6 Kelvin
2005	2 to 3 Kelvin	approx. 1 Kelvin

The reason the values of the construction year 1995 are higher than those of 1965 can be explained as follows: The continuous concrete slab represents a very efficient thermal bridge. Hence, the difference between the concrete slab and the better insulated construction of the wall is greater than the difference between the concrete slab and the poorly insulated wall of construction year 1965. Normally, the value of construction year 1995 would be lower, as for this category a thermal separation can be presumed.

Furthermore, the results show that for indoor thermography, there is no significant deviation of the average values, irrespective of the suitability of the outside climate. The difference between the well-suited and unsuitable climatic boundary conditions is in the same range.

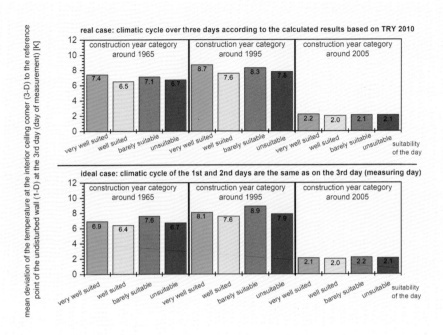

Fig. 5-10 Maximum temperature variations of the interior ceiling corner (3-D) compared to the undisturbed construction (1-D).

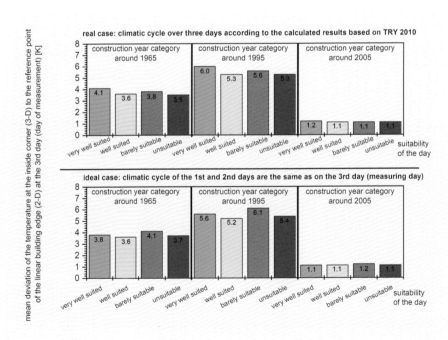

Fig. 5-11 Maximum temperature variations of the interior ceiling corner (3-D) compared to the linear building edge (2-D).

Accuracy of Temperatures Indicated in Thermograms

5.1.3 The Effect of Defective Points in Thermal Insulation

Within the calculations framework, the impact of defective locations on the interior surface temperature and the exterior surface temperature has been studied. Fig. 5-12 represents a typical result of the calculations. The diagrams show that a void (filled with air or mortar) within a well-insulated construction of the corner of a building can be detected by means of outdoor thermography rather than by means of indoor thermography. The temperature deviations on the interior are within the measurement inaccuracy of thermography cameras, so localization indoors is almost impossible.

Fig. 5-12 shows the evaluated thermography results of a 50 mm wide void, filled with mortar or air. It can be seen that on the exterior, the voids stand out much more than on the interior.

In order to locate voids by means of outdoor thermography with sufficient accuracy, the voids must be a certain size. As a rule, voids greater than about 10 mm are detectable, since they exhibit a higher temperature on the exterior due to the thermal bridge effect, compared to the undisturbed construction.

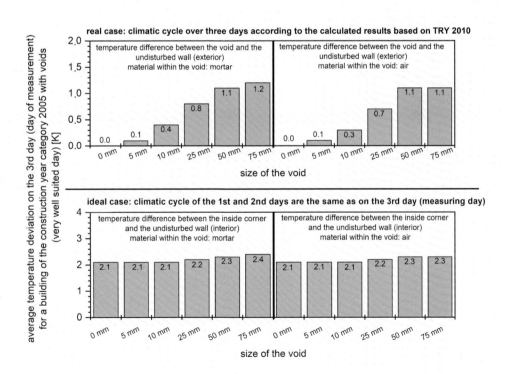

Fig. 5-12 Maximum deviations / average deviations of the temperatures at the interior ceiling corner (3-D, *bottom*) and on the exterior (*top*), in order to determine where voids within the thermal insulation composite system are detectable (evaluation of the thermography results of a 50 mm wide void, filled with air or mortar). Due to higher contrast, the detectability of voids within the thermal insulation composite system is better on the exterior.

5.2   Effects of Wrongly Assumed Emissivity and Background Radiation on the Measurement Results

The physical basics for the measurement of surface temperature were discussed in Chapter 2.2.2. In this section the possible deviation ranges of the detected temperatures for different emissivities and ambient radiations are shown with the aid of an example of an indoor thermography measurement and an outdoor thermography measurement. Subsequently, this information will be used to classify the impact of the boundary conditions mentioned above on the temperatures indicated in the measuring results.

5.2.1   Example of Indoor Thermography

For the example of indoor thermography, the case of a 3-D thermal bridge discussed in Section 5.1 and as example in Chapter 7.1.1 is used. By means of a contact thermometer, the actual temperature of the construction surface was determined (see Fig. 5-13). The surface temperature θ_{si} of the ceiling corner at the time of the thermography measurement was 15.6°C.

Fig. 5-13 Measuring the surface temperature with the aid of a contact thermometer. Displayed measuring value: θsi = 15.6°C.

The spectrum of the displayed temperature changes with the variation of emissivity and ambient radiation temperature, which can be adjusted at the thermography camera. In Fig. 5-14 the spectra are shown for the interior ceiling corner. In addition, the behavior of the measuring value for altered emissivity is indicated.

In most cases for indoor thermography, the ambient radiation is very close to the interior temperature. The reason for this is that in general, the interior walls and the furniture exhibit this temperature. When using a common emissivity of material surfaces, as is done in Fig. 5-14 (ε is presumed to be about 0.90 to 0.98), a deviation from the actual surface temperature of $\Delta\theta < \pm 0.2$ Kelvin can occur. These measurement errors are within the range of the measurement inaccuracy of the thermography systems.

When presuming ambient radiation temperatures that are far too low ($\theta_U = \pm0°C$) or too high ($\theta_U = +40°C$), measurement errors of $\Delta\theta < \pm3$ Kelvin can occur. Only for the combination of implausibly (erroneous) high or low ambient radiation temperatures with a completely

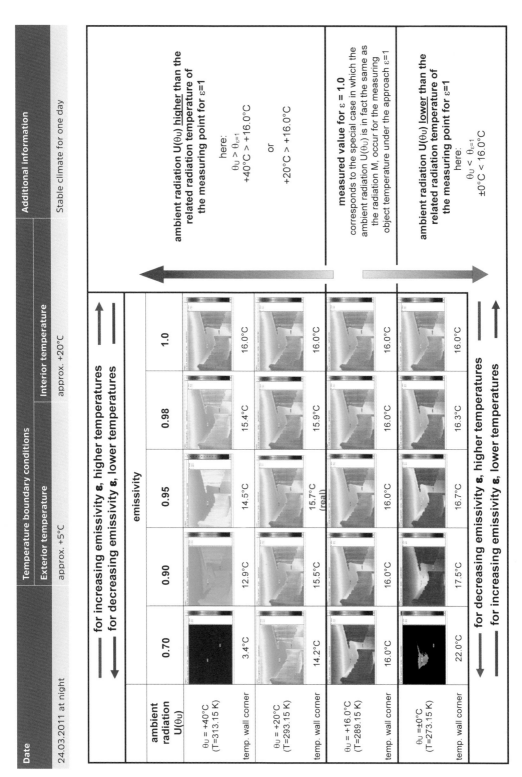

Fig. 5-14 Spectrum of the displayed surface temperature of the interior ceiling corner for different emissivity and ambient radiation temperatures; actual temperature θ_{si} in the corner is 15.6–15.7°C.

wrong emissivity (in the example: ε = 0.70), the deviation of the temperatures detected by means of indoor thermography can reach into the two-digit range (in the example, see Fig. 5-14: −12 Kelvin or +6 Kelvin).

5.2.2 Example of Outdoor Thermography

An outdoor thermography was performed on the thermal bridge discussed in Chapter 7.1.1. The surface temperature at the interior ceiling corner was measured by means of a contact thermometer. The surface temperature θ_{si} of the corner at the time of the thermography measurement was about 11.2°C (see Fig. 5-15).

At the time of the thermography measurement the sky was clear. For such boundary conditions, the ambient radiation temperature is usually significantly lower than the exterior temperature. By means of the image showing the viewing range of the corner that is to be thermographed, the ambient radiation temperature can be determined according to the method described in Fig. 2-25.

For the situation depicted in Fig. 5-17, the ambient radiation temperature θ_U was determined to be about −20°C. For common emissivity of the material surfaces (ε = 0.90–0.98) and an estimated ambient radiation temperature $\Delta\theta_U$ of ±20 K compared to the real ambient radiation temperature θ_U of −20°C, an effective range of the displayed surface temperatures of $\Delta\theta < \pm 2$ Kelvin compared to the actual surface temperature can be reached.

The impact of the ambient radiation temperature on the measuring value increases with smaller emissivity. Thus, for additional uncertainties concerning the determination of the ambient radiation temperatures, the deviations of the displayed surface temperature from the actual surface temperature can easily reach double-digit values. People who are familiar with thermography know the challenge of measuring metal surfaces. However, small emissivity (ε ≈ 0.7) can also occur when measuring coated windows by means of thermography or when measuring at a flat angle (see Chapter 2.1.2). In such cases, the measuring accuracy of thermograms should be called into question.

Fig. 5-15 Exterior perspective of the building, measuring of the surface temperature in the exterior ceiling corner (left side of the balcony on the upper floor) by means of a contact thermometer; displayed measuring value: θ_{se} = 11.1°C.

Accuracy of Temperatures Indicated in Thermograms

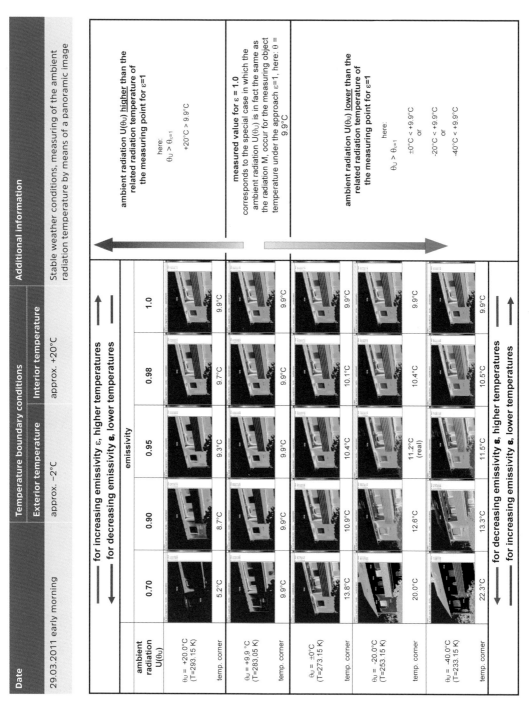

Date	Temperature boundary conditions		Additional information
	Exterior temperature	Interior temperature	
29.03.2011 early morning	approx. –2°C	approx. +20°C	Stable weather conditions, measuring of the ambient radiation temperature by means of a panoramic image

for increasing emissivity ε, higher temperatures
for decreasing emissivity ε, lower temperatures

ambient radiation U(θu)	emissivity				
	0.70	0.90	0.95	0.98	1.0
θu = +20.0°C (T=293.15 K)					
temp. corner	5.2°C	8.7°C	9.3°C	9.7°C	9.9°C
θu = +9.9°C (T=283.05 K)					
temp. corner	9.9°C	9.9°C	9.9°C	9.9°C	9.9°C
θu = ±0°C (T=273.15 K)					
temp. corner	13.8°C	10.9°C	10.4°C	10.1°C	9.9°C
θu = -20.0°C (T=253.15 K)					
temp. corner	20.0°C	12.6°C	11.2°C (real)	10.4°C	9.9°C
θu = -40.0°C (T=233.15 K)					
temp. corner	22.3°C	13.3°C	11.5°C	10.5°C	9.9°C

for decreasing emissivity ε, higher temperatures
for increasing emissivity ε, lower temperatures

ambient radiation U(θU) <u>higher</u> than the related radiation temperature of the measuring point for ε=1

here:
$\theta_U > \theta_{\varepsilon=1}$
+20°C > 9.9°C

measured value for ε = 1.0 corresponds to the special case in which the ambient radiation U(θU) is in fact the same as the radiation M, occur for the measuring object temperature under the approach ε=1, here: θ = 9.9°C

ambient radiation U(θU) <u>lower</u> than the related radiation temperature of the measuring point for ε=1

here:
$\theta_U > \theta_{\varepsilon=1}$
±0°C < +9.9°C
or
-20°C < +9.9°C
or
-40°C < +9.9°C

Fig. 5-16 Spectrum of the displayed surface temperature of the exterior wall corner for different emissivity and ambient radiation temperatures; actual temperature in the corner: θ_{se} = 11.1–11.2°C.

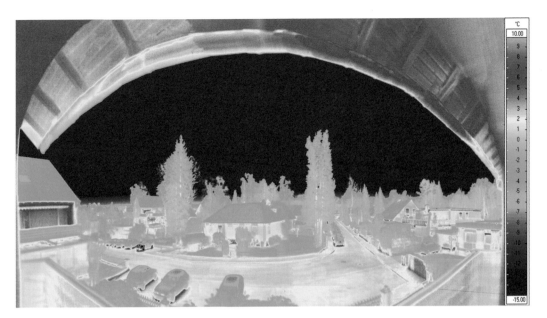

Fig. 5-17 Thermogram used to determine the ambient radiation temperature, half-space image taken in front of the wall that is to be measured by means of thermography; ambient radiation temperature: θu ≈ −20.0°C. The evaluation point of the thermograms depicted in Fig. 5-16 is on the very right edge of this photograph.

Accuracy of Temperatures Indicated in Thermograms

6 Application of Thermography in Civil Engineering

6.1 Basic Principles and Measurement Conditions

The objective of thermography is to create snapshots that visualize the surface temperature or the surface temperature distribution of a particular moment. By means of the surface temperature distribution, thermomechanical irregularities (such as those due to thermal bridges), differing moisture contents, or air leaks in the building envelope can be localized.

In order to get an overall picture of the building, thermography measurement should always be conducted from both indoors and outdoors. The following sections only focus on passive thermography, which is commonly used in practice. The application of active thermography techniques, such as impulse thermography or pulsed-phase thermography, is comprehensively discussed in the *Bauphysik-Kalender 2004* [44]. Interesting fields of application of active thermography in civil engineering are discussed in [42] and [43].

6.1.1 Measurement Conditions for Outdoor and Indoor Thermography

In order to localize thermal bridges, thermography measurement is often conducted from outdoors. In most cases this is of advantage compared to indoor thermography, since a wider field of view can be covered with one image. For detection of thermomechanical irregularities, a sufficient temperature difference between indoors and outdoors must be available. In specialist literature, a minimum temperature difference of at least 15 Kelvin is recommended. The higher the temperature difference, the better. Presuming an average interior temperature of about +20°C, thermography measurement should not be conducted for exterior temperatures higher than about +5°C. Of course, thermography measurement can also be carried out for warmer outside climates. However, for smaller temperature differences, potential irregularities are much more difficult to identify, if at all.

Since thermography measurements always represent a snapshot of transient conditions, the temperature difference stated above must also be present before the measuring. Preferably, the temperature differences should be constant and last as long as possible. In general, 12 to 24 hours are considered sufficient. If the exterior temperature varies significantly throughout the day, the thermography measurement results should be questioned. In Germany, sufficient temperature differences are usually given for the period between October and April.

Thermograms: Outdoor Thermography

When generating outdoor thermograms, special attention must be paid to the effects of solar radiation. The measuring object must not be influenced by direct solar radiation for the time of the measurement and a sufficient period beforehand. The thermal storage effects of surfaces that have been directly heated by solar radiation can lead to the misinterpretation of measurements. Furthermore, direct solar radiation is reflected and influences the optical beam path of the IR camera (see Chapter 2.2, Fig. 2-24). Shadows caused by direct solar radiation, as well as diffuse solar radiation (strongly scattered, not causing shadows), can also lead to wrong measuring results.

In addition, the radiative exchange of the measuring object surface with the environment must also be considered. On starry nights, the surface temperatures can be lower than the exterior temperature, as the sky background temperature can be as low as −40°C, which causes a higher influence of the radiative exchange with the building surface (influence of the ambient radiation temperature θ_U; see Chapter 2.2). For overcast skies, the radiative exchange of the surfaces mainly takes place between the building surface and the clouds that are not quite as cold as the sky and therefore, the cooling of the building surface is less.

The surface temperatures of construction parts are strongly influenced by the effects of convective heat transfer. Therefore, wind conditions must also be considered when conducting thermography measurements. Depending on the type of the problem, it must be determined if the wind or the high fluctuation of the wind intensity has an impact on the objective of the measurement. Reference values for maximum wind velocities are v = 2 m/s [40] and up to 6.7 m/s (15 mph) [2].

Further influencing factors on the surface temperature are condensation and rainwater on the building surface. Condensation can appear on the surface of very well insulated exterior wall constructions or windows. This is an indication of very good thermal insulation. However, the evaporation of water at the surface of moist objects leads to a cooling of the object due to the extraction of heat energy needed for the evaporation process.

Fig. 6-1 Examples of how it should not be done: Impacts of shadows and direct solar radiation on thermography images.

Application of Thermography in Civil Engineering

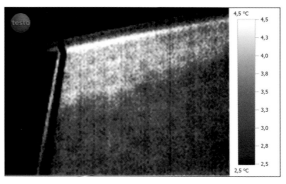

Fig. 6-2 Cooling of an exterior wall caused by moisture on the surface.

Possible effects of temporary humidification or moisture penetration on thermograms are shown in Fig. 6-2. The exterior wall surface has been humidified by soft rain (fog). Under the roof overhang, the wall was protected from the rain. Thus, the expected temperature distribution has been falsified by the moisture penetration of the exterior wall, since in reality, the upper floor is heated and the jamb wall and the attic are not. In the thermogram, the unheated jamb wall should appear cooler than the heated upper floor. But due to the moisture penetration, the thermogram displays a different image.

Based on the boundary conditions mentioned above, the best time for outdoor thermography measurements is in the early morning hours before sunrise. The early morning is to be preferred to nighttime, because normally the lowest temperatures are present in the morning due to overnight cooling. By extension, winter is the best season for common construction thermography.

Thermograms: Indoor Thermography

When generating thermograms indoors, the captured area of the exterior wall surface is much smaller compared to outdoor thermography – though in most cases thermal bridge phenomena are depicted in greater detail due to better temperature resolution (see Fig. 6-3). The view from inside is often restricted by furniture and other fitments; thus, more effort may be necessary to prepare for thermography measuring.

The advantages of indoor thermography are: (1) the air temperature in residential buildings remains relatively constant over a longer period; and (2) the climatic influences that have to be considered for outdoor thermography have a marginal impact and thus only need to be considered in weakened form. Therefore, in most cases, indoor thermography allows for thermography measurement even during the day. Moreover, it can be said from experience that for developed attics with rear-ventilated roofs or constructions with rear-ventilated exterior walls, sound evaluation is only possible by means of indoor thermography. Thus, in most cases, indoor thermography has a higher informative value and is preferably used for concrete evaluation.

6.2 Localization of Thermal Bridges by Means of Thermography

6.2.1 Systematics of Thermal Bridges

Thermal bridges are local areas within room-enclosing construction parts that exhibit an increased heat flow from the warmer side to the colder side. Thermal bridges can lead to the following effects:

- increased heat loss
- decreased inside surface temperature at the thermal bridge compared to the surface of the undisturbed construction, which may cause the accumulation of condensation and mold infestation
- decreased interior surface temperature, which affects thermal comfort.

With regard to building physics, the following thermal bridges are to be distinguished.

6.2.2 Material-Related Thermal Bridges

Material-related thermal bridges exist if there is a significant difference in thermal conductivity between the materials. The temperature distribution in Fig. 6-3 shows that the difference of the surface temperature between the thermal bridge and the undisturbed construction is considerably higher on the interior than on the exterior. This is one of the reasons why, in most cases, indoor thermography allows for a more detailed evaluation of thermal bridges.

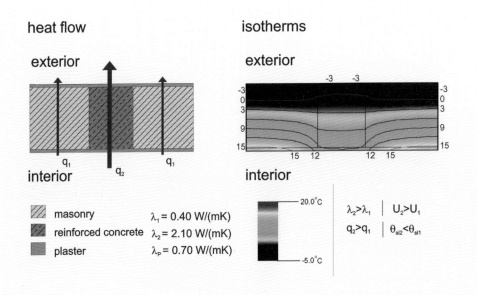

Fig. 6-3 Definition of a material-related thermal bridge.

6.2.3 Geometric Thermal Bridges

Geometric thermal bridges are characterized by a small heat-absorbent surface A_i on the interior compared to a much larger heat-emitting surface A_e on the exterior. In the corner of a construction, this leads to a higher total heat transition and thus, to a lower surface temperature in the disturbed corner area. This effect is most powerful for three-dimensional examination of the thermal bridge effect (for example, an interior corner between the exterior wall and interior ceiling). As has already been observed, in most cases indoor thermography allows for a more detailed evaluation of thermal bridges due to better temperature resolution. Combinations of material-related and geometrical thermal bridges are also possible (for example, the overhanging concrete slab of a balcony going through a masonry wall).

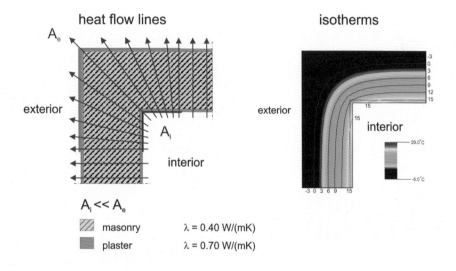

Fig. 6-4 Definition of a geometrical thermal bridge.

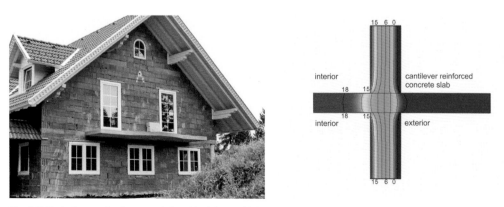

Fig. 6-5 Typical example of a potentially critical thermal bridge: cantilevered reinforced concrete slab of a balcony, in this case without correct thermal isolation.

Based on the definitions stated above, it can be seen that thermal bridges are unavoidable and thus exist in every building. However, by means of proper technical planning and implementation, thermal bridges within a construction can be reduced to a minimum. This means that the negative effects of thermal bridges, such as increased heat loss and low interior surface temperatures are reduced to an uncritical minimum (for verification see Section 6.2.5).

6.2.4 Typical Thermograms of Thermal Bridges without Damage

Due to the available camera technology, the problem of not knowing the exact emissivity of the surfaces and the uncertain impact of ambient radiation, it is not possible to exactly determine the absolute temperature by means of construction thermography in many cases. However, it is usually not necessary to exactly determine the temperature, since in practice well-detectable temperature differences between the building part surfaces allow for an evaluation of the construction.

Thus, the main objective of detecting thermal bridges by means of construction thermography is to localize anomalies within the construction characterized by significant temperature differences in the thermogram. Then, based on professional expertise and experience, the thermography expert has to decide whether the measured temperature differences are normal or conspicuous for the existing boundary conditions.

Since temperature differences caused by thermal bridges also occur within faultless constructions (see section above), the following sections are to show thermograms of constructions that are free from defects.

Corner of a Wall: Example no. 1

Every corner of a wall exhibits a certain decrease of the interior surface temperature. This doesn't automatically indicate a defect (see Section 6.2.1).

Within the framework of thermography, it is common practice to compare the temperature of the corner area with the temperature of the undisturbed construction. Constructions that are free from defects will only exhibit marginal temperature differences ($\Delta\theta_{si}$ approx. 2–3 K), and the surface temperature near the corner increases promptly. Furthermore, the temperatures of the surrounding undisturbed constructions are rather uniform and constant.

For greater temperature differences (without any logical explanation), this area might represent a defect and should be further examined according to the approach discussed in Section 6.2.5.

Application of Thermography in Civil Engineering

Date	Temperature boundary conditions		Additional information
	Exterior temperature	Interior temperature	
13.11.2005 at night	approx. +2°C	approx. +22°C	./.

Fig. 6-6 Corner of a building with a thermal insulation composite system and free from defects (thickness of insulation layer: 10 cm, WLG 040, according to German classification).

 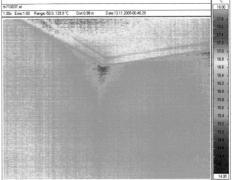

Fig. 6-7 Interior view of the defect-free corner of the exterior wall depicted in Fig. 6-6.

Corner of a Wall: Example no. 2

Fig. 6-8 shows another example of an uncritical corner of an exterior wall. The temperature in the corner is about 3 Kelvin lower than the temperature of the undisturbed wall. However, it increases promptly with the distance from the corner.

Date	Temperature boundary conditions		Additional information
	Exterior temperature	Interior temperature	
27.01.2005 evening	approx. 0°C	approx. +22°C	./.

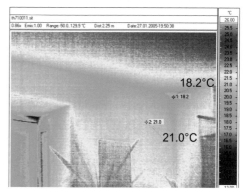

Fig. 6-8 Corner of a defect-free exterior wall with corresponding thermogram.

Windows: Example no. 1

Windows are generally the part of the construction that has a lower thermal insulation than the exterior walls, both for old and new buildings. Especially the window frame and the thermal problematic interconnection between the glazing and the frame stand out in thermograms (such as in the indoor thermogram depicted in Fig. 6-10). Furthermore, air on the exterior of the windows heats up and accumulates under the edge of the wall, which for relatively windless conditions leads to a typical warming of the lintel.

Date	Temperature boundary conditions		Additional information
	Exterior temperature	Interior temperature	
13.11.2005 at night	approx. +2°C	approx. +22°C	./.

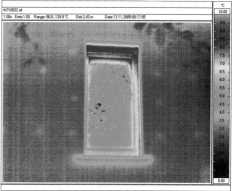

Fig. 6-9 Exterior view of a wooden window construction and its corresponding thermogram: in the thermogram the mounting screws of the thermal insulation composite system become visible.

Application of Thermography in Civil Engineering

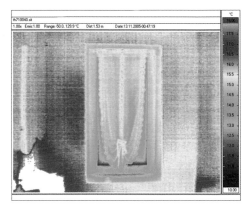

Fig. 6-10 Indoor thermography of the window construction from Fig. 6-9: the interconnection between the glazing and the window frame stands out as a cold area.

At this point it has to be mentioned that in this case, the significant heating of the lintel and the top of the window frame shown in Fig. 6-9 is due to the bathroom window having been open until about half an hour before the beginning of the thermography measurement. Thus, in practice, it is very important to consider the boundary conditions at the time of the thermography measurement and prior to it. Furthermore, some irregularities can be explained by querying the user or occupant, as in the example of Fig. 6-9.

Windows: Example no. 2

Modern office buildings frequently feature extensive window areas. Fig. 6-11 shows a thermogram of such a building. As above, the (wooden) window frames have a lower thermal protection than the surrounding walls and thus exhibit higher surface temperatures in the outdoor thermogram. Furthermore, the typical warm air area around the window lintels that can be seen in almost every thermogram is also visible.

Date	Temperature boundary conditions		Additional information
	Exterior temperature	Interior temperature	
10.03.2005 at night	approx. 0°C	approx. +20°C	The thermogram is an assembly of three single thermograms; reflections of the clouds and the sky are visible in the thermogram

Fig. 6-11 Extensive window construction in an office building: the common thermography image shows windows that are thermally free from defects.

Thermally Isolated Construction Parts

The thermal isolation of construction parts from the outdoor area is especially important for building parts with high temperatures (isolation of heated screeds) or building parts with high thermal conductivity (thermal isolation of steel beams). Fig. 6-12 shows an example of an ideally thermally isolated threshold of a terrace door. In the thermogram, the effectiveness shows in the steep decrease of the temperature at the threshold. An example of a poorly implemented thermal isolation of a steel beam is shown in Chapter 7.1.3.

Date	Temperature boundary conditions		Additional information
	Exterior temperature	Interior temperature	
27.01.2005 evening	approx. 0°C	approx. +23°C	Floor heating indoors

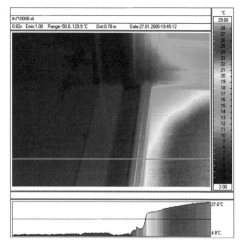

Fig. 6-12 Thermally isolated threshold of a terrace door.

Exterior Wall Construction of a High-Rise Building

With the aid of imaging thermography, exterior wall constructions can quickly be examined with regard to their thermal behavior. The following figure shows the facade construction of a high-rise building. The exterior wall surface exhibits a uniform temperature level throughout, without any defects. Local temperature peaks (so-called hot spots) indicate windows that were open during the thermography measurement.

Date	Temperature boundary conditions		Additional information
	Exterior temperature	Interior temperature	
26.02.2005 noon	approx. 0°C	approx. 22°C	low cloud cover

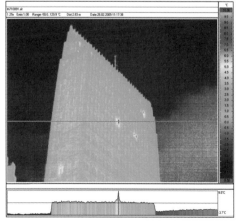

Fig. 6-13 Facade of a high-rise building; hot spots indicate open windows.

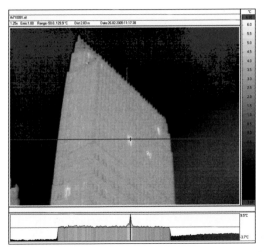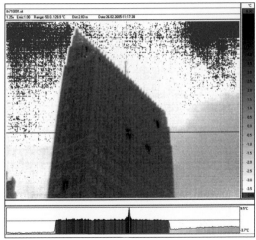

Fig. 6-14 The same thermogram as in Fig. 6-13: By means of evaluation software interesting areas can be emphasized.

6.2.5 Computational Approaches for the Examination of Thermal Bridges

With the aid of thermography, the effects of thermal bridges can be visualized and weak spots can be detected. Thus, thermography is an important nondestructive tool for the evaluation of building constructions. Regarding thermal bridges, the main purpose of thermography is to help evaluate if the thermal bridge causes any damage or not. Hence, in many cases the question is whether the cause of moisture damage or mold formation should be ascribed to faulty construction or to user behavior.

In addition to thermography measurement that is used to localize the defect area for further examination (for example, taking representative samples or opening the construction), a computational examination should be carried out. As a matter of course, it is necessary to know the structure of the construction and its material properties.

Preventing Low Interior Surface Temperatures and Mold Formation

Mold formation occurs when mold spores, which are always present in room air, meet favorable living conditions. Studies on mold formation have shown that the crucial factor in mold formation is the surface moisture of the construction parts. It can therefore be assumed that if the room air near the construction surface cools down so much that the relative air humidity is 80%, mold formation can occur (according to current studies, under certain circumstances mold formation can already occur directly on the component surface at a relative air humidity of under 70% [37]).

A sufficient nutrient supply is another important factor for mold formation. In this context, severe pollution can considerably encourage mold formation. Further parameters, such as pH-value, oxygen content, CO_2 content, and light have only minor impact.

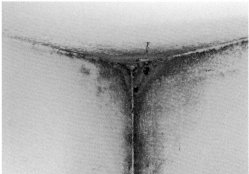

Fig. 6-15 Extreme mold formation on the inside corner between exterior walls and the ceiling and along the window soffit.

Since in practice, material-related and geometrical thermal bridges cannot always be avoided, it must be defined at which point a thermal bridge causes defects in the construction. In this context the DIN EN ISO 13788 [15] and DIN 4108-2 [12] provide relevant instructions. The DIN EN ISO 13788 suggests an approach on a monthly basis and a temperature factor f_{Rsi} of the construction that is not to be undercut by the interior surface temperature. According to DIN 4108-2, a thermal bridge is considered destructive if under certain stationary boundary conditions, the temperature factor f_{Rsi} [./] undercuts the result of equation (32).

$$f_{Rsi} = \frac{\theta_{si} - \theta_e}{\theta_i - \theta_e} \geq 0.70 \qquad (32)$$

with:

θ_{si} inside surface temperature [°C]

θ_i interior temperature [°C]

θ_e exterior temperature [°C]

The boundary conditions presumed for the approach of DIN 4108-2 [12] are as follows:

Tab. 6-1 Climate boundary conditions used for the verification of the minimum thermal insulation of thermal bridges in order to prevent surface condensate and mold formation.

Exterior temperature:	$\theta_e = -5°C$	(average of a five-day cold period in Germany)
Interior temperature: rel. humidity:	$\theta_i = +20°C$ $\phi = 50\%$	(common air temperature and relative humidity of a "typical" living room in use)

For thermal bridges that are not in the living area, the DIN 4108-2 provides the following temperature boundary (outside) conditions:

Basement, ground: $\qquad\qquad\qquad\qquad\qquad\qquad\qquad\qquad$ $\theta_e = +10°C$
Unheated buffer zone (such as an unheated staircase): $\quad$ $\theta_e = +10°C$
Unheated attic or underground car garage: $\qquad\qquad\qquad$ $\theta_e = -5°C$

According to DIN 4108-2 the heat transmission resistance is determined as follows:

Outdoors: $\qquad\qquad\qquad\quad$ $R_{se} = 0.04 \text{ m}^2\text{K/W} \rightarrow$ inverse: $h_{se} = 25$ W/(m^{2}K)
Indoors, heated rooms: $\quad$ $R_{si} = 0.25 \text{ m}^2\text{K/W} \rightarrow$ inverse: $h_{si} = 4$ W/(m^{2}K)
Indoors, unheated rooms: $R_{si} = 0.17 \text{ m}^2\text{K/W} \rightarrow$ inverse: $h_{si} = 5.882$ W/(m^{2}K)

With regard to the indoor climates stated above, the critical inside surface temperature to prevent mold formation is $\theta_{si,cr} = +12.6°C$. Thus, for the computational evaluation of thermal bridges regarding mold formation, it must be determined if the surface temperature of the construction part exceeds the critical value. The requirement of $f_{Rsi} \geq 0.70$ stated in equation (32) can be explained by inserting the standard boundary conditions according to DIN 4108-2 and the critical surface temperature for mold formation:

$$f_{Rsi} = \frac{\theta_{si} - \theta_e}{\theta_i - \theta_e} = \frac{12.6 - (-5)}{20 - (-5)} = 0.70 \qquad\qquad (33)$$

with:
θ_{si} $\qquad$ inside surface temperature, here $\theta_{si,cr} = 12.6°C$
θ_i $\qquad$ interior temperature, according to DIN 4108-2
θ_e $\qquad$ exterior temperature, according to DIN 4108-2

In the event the actual climate boundary conditions differ from the standard boundary conditions (for example, a swimming pool or an air-conditioned building), the requirements have to be adjusted and the critical surface temperature or f_{Rsi} must be redetermined. An example of a computational examination is discussed in Chapter 7.1.6.

In most cases, a direct metrological examination of the approach stated in equation (33) is not possible, since in practice, boundary conditions are not stationary and the presumed room climate is not necessarily given. However, for long periods of consistent weather and indoor boundary conditions, equation (33) can be used to roughly estimate the correlation between the boundary conditions and the measuring values for the existing temperatures.

Application of Thermography in Civil Engineering

The critical surface temperature for given boundary conditions can be determined by rearranging equation (33) as follows:

$$f_{R_{si}} \cdot (\theta_i - \theta_e) + \theta_e = \theta_{si} \tag{34}$$

This equation can then be used to roughly evaluate the measuring values.

Example: The exterior temperature has been nearly constant at about +3°C for a long period and the interior surface temperature was determined to be 14°C for a nearly constant heating of about 22°C. In order to estimate if this temperature is to be considered critical, the approach of equation (34) can be used:

$$0.7 \cdot (22°C - 3°C) + 3°C = 16.3°C \tag{35}$$

Considering that the boundary conditions have been nearly stationary, an interior surface temperature of 14°C is too low. Hence, this area of the exterior wall should be further examined.

As mentioned above, the actual instationary conditions can only be represented in simplified terms by the stationary verification procedure according to DIN 4108-2. For further analysis regarding building physics, combined hygrometric simulations can be used to calculate the instationary process of heat, moisture, and salt transport through the building parts. Software programs commonly used for such calculations are DELPHIN [9] and WUFI [45]. Further calculations can also be used to forecast the risk of mold formation (such as WUFI Bio [46]). Further publications have been prepared, for example, by the WTA working group [47]. These methods allow for quite realistic forecasts on the actual behavior of the construction.

Energetic Evaluation: Determination of Heat Loss

Within the framework of energy-saving heat protection (energy-saving regulations – in Germany the EnEV, since November 2020 the GEG [49]), the knowledge of additional heat loss caused by thermal bridges is of great importance. With regard to the calculations, the transmission heat loss H_T caused by thermal bridges can generally be considered by means of the thermal bridge loss coefficient:

It is differentiated between: ψ (Psi), Linear Transmittance, indicating the heat loss due to linear thermal bridges per meter and Kelvin, [W/(mK)] (such as the edges of the building, window soffits, door soffits, overhanging constructions, connections between wall and ceiling); and χ (Chi), Point Transmittance, indicating the heat loss due to point thermal bridges per Kelvin [W/K] (such as mounting elements within the facade).

The indication of the thermal bridge loss coefficient refers to a reference plane (area), which is usually stated as the interior length of the wall section (translation to the exterior length of the wall section also is possible).

The thermal bridge loss coefficients are determined by means of sophisticated numerical programs or by using thermal bridge atlases, which list the coefficients for common constructions. A quantification of the thermal bridge loss coefficients by means of thermography is not possible. However, thermography allows for a visual interpretation of the thermal bridge phenomena.

For the application within the framework of the energy-saving heat protection, the general equation given in Fig. 6-16 is modified by so-called temperature correlation factors F. For further literature on thermal bridges and their relevance for civil engineering see for example [6], [26], and [28].

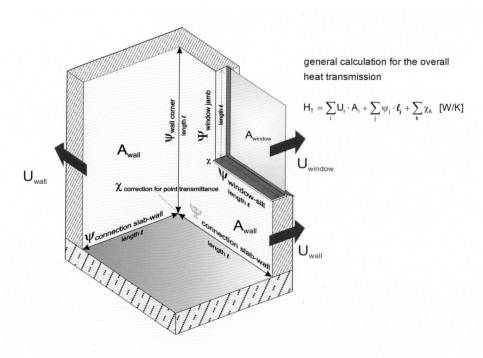

general calculation for the overall heat transmission

$$H_T = \sum_i U_i \cdot A_i + \sum_j \psi_j \cdot \ell_j + \sum_k \chi_k \quad [W/K]$$

Fig. 6-16 Application of thermal bridge loss coefficients taking account of the heat loss caused by thermal bridges, illustrated by means of a wall-to-ceiling connection.

7   Practical Examples

7.1 Thermal Bridges

7.1.1 Exterior Wall Corner with Overhanging Concrete Ceiling

In an apartment building (construction year 1965) an accumulation of condensation and beginning of mold formation was found in the corner of an exterior wall in a workroom. Within the framework of sourcing the problem, a thermography analysis of this area of the wall was conducted. The photograph taken from indoors shows the corner of the wall of the workroom, which adjoins a balcony on the exterior.

The thermal image taken from indoors shows a significant cooling of the wall areas close to the ceiling and the neighboring ceiling areas (the difference to the undisturbed wall construction is about 5 Kelvin). The reason for this significant cooling on the interior wall surface is the geometric design of the construction in the corner region. In addition to the unavoidable geometric thermal bridge of the corner region, the concrete ceiling is not thermally isolated from the adjoining exterior wall (see Fig. 7-3).

Date of thermal imaging	Temperature boundary conditions		Additional information
	Exterior temperature	Interior temperature	
22.12.2004 evening	approx. −1°C	approx. +22°C	Indoor shot; stable cold weather conditions over a long period of time

Fig. 7-1 Exterior wall corner region with beginning mold.

There is a geometric thermal bridge (due to the corner) as well as a material-related thermal bridge (due to the traversing ceiling and lintel). It has to be mentioned that at the time of the building's construction, the isolation of overhanging construction elements wasn't state-of-the-art, and the standards for thermal insulation are not comparable to current standards.

The renters were advised to move the shelf further away from the corner region (see Fig. 7-1). Through this measure, the heat transmission resistance on the inside was decreased, which led to a better heating of the interior corner region.

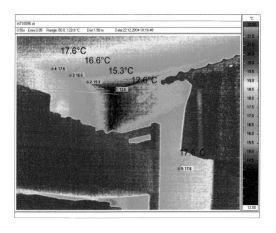 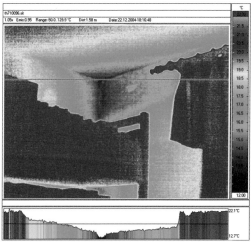

Fig. 7-2 Thermal image of the corner region. With the aid of evaluation software it is possible to display single reference temperatures along the ceiling-wall crossover (*left*) or a permanent temperature distribution along a reference line (*right*).

Fig. 7-3 Exterior view of the balcony with the traversing concrete ceiling.

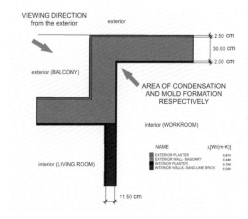

Fig. 7-4 Geometry and design of the corner, horizontal section below the overhanging ceiling.

The corner area in Fig. 7-1 was thermographed again in November 2016 and shown in Fig. 7-5. In addition to the conspicuously clearer image due to the higher geometric resolution (1024 × 768 pixels compared to the original image's 320 × 240 pixels in Fig. 7-2), almost the same temperature differences are determined at the evaluation points.

Date of thermal imaging	Temperature boundary conditions		Additional information
	Exterior temperature	Interior temperature	
27.11.2016 at noon	approx. +3°C	approx. +22°C	a technically better camera was used for this thermogram

Fig. 7-5 Thermogram of the wall-corner area already shown in Fig. 7-2. The difference between the 3D-lxel and the almost undisturbed wall area (1D) is similar to the thermogram in Fig. 7-2 from 2004 (about 5 Kelvin), thermogram shot with a camera with a significantly higher geometric resolution.

7.1.2   Uninsulated Lintel

A typical material-related thermal bridge exists when there are construction elements with different thermal conductivities (see Chapter 6.2.1). A typical example of this is an uninsulated concrete window lintel. Such a construction element is shown in Fig. 7-6 as well as in Fig. 7-7 as a thermal image.

Date	Temperature boundary conditions		Additional information
	Exterior temperature	Interior temperature	
13.11.2004 at night	approx. 0°C	approx. +22°C	Outdoor shot, stable climate conditions (exterior air temperatures around 0°C during the preceding four days)

Fig. 7-6 Masonry construction with uninsulated lintel (reinforced concrete); (*right*) photograph taken during the construction process (approx. 1986) showing the position of the window lintel made of concrete.

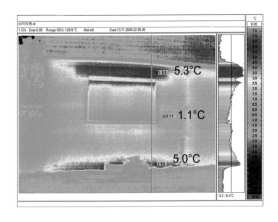

Fig. 7-7 Thermal image with vertical temperature distribution. When the thermography measurement was conducted (at night), the blinds of the window were closed. Waste heat is shown emanating from the basement, where the house's heating system is located.

The thermal image clearly shows the impact of the thermal bridge of the concrete window lintel through a significantly elevated surface temperature. Thus, the surface temperature of the uninsulated lintel is about θ_{se} = 5.3°C, whereas the surface temperature of the nearly undisturbed wall is θ_{se} = 1.1°C.

Another thermal bridge can be seen in the area between the basement and the ground floor. The oil-fired heating system of this house is located in the basement and produces a remarkable amount of waste heat, which allows for an interpretation of the high surface temperatures.

With the aid of these images, the owner of the building was shown the thermal conditions of the building and its weak spots. A significant improvement of this situation is only possible through installing thermal insulation. Needless to say, it is important to consider the cost-effectiveness of this measure: the energy savings and the cost of the insulation and its installation must be compared. However, if there are renovation works or structural alteration works planned, the additional expenses for thermal insulation – which are minor in most cases – should be considered. It should also be mentioned that there is legislation with measures to improve thermal insulation standards in any event (EnEV=energy saving regulation of the German government; since November 2020, this is replaced by the GEG [Gebäude Energie Gesetz=Building energy law] [49]).

In addition, Fig. 7-8 shows a thermogram of the same section of the building taken at a different time with a state-of-the-art detector format of 1024 × 768 infrared pixels (camera: InfraTec VarioCam HD). With the higher resolution, significantly sharper and more detailed thermograms with more information content can be achieved. Thus, the dimensions of the masonry used can also be seen in the thermogram (see in particular Fig. 7-6, *right*).

Date	Temperature boundary conditions		Additional information
	Exterior temperature	Interior temperature	
22.02.2015 at night	approx. +2°C	approx. +22°C	

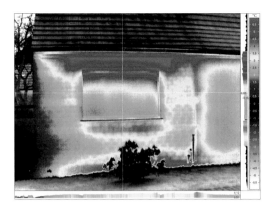

Fig. 7-8 Thermogram with the same content as in Fig. 7-7, recorded here with an infrared camera with significantly better resolution and associated higher detail density.

7.1.3 Steel Beams without Thermal Isolation Traversing the Facade

With regard to thermal bridges, building elements consisting of materials with high thermal conductivity can be especially critical. The next example shows a steel beam that connects a very warm interior (indoor swimming pool) with the exterior without any insulation. At the point where the beam traverses the building envelope, air leaks and interruptions of the thermal insulation layer were found. The impact of the thermal bridge in the region of the beam's transition is illustrated in the thermal image (see Fig. 7-9).

Such thermal bridges are usually thermally separated by an auxiliary construction in order to interrupt or decrease the thermal conduction. A suitable construction to connect two steel profiles is shown in Fig. 7-10.

Date	Temperature boundary conditions		Additional information
	Exterior temperature	Interior temperature	
01.03.2002 morning	approx. +2°C	approx. +32°C	High interior temperature due to indoor swimming pool

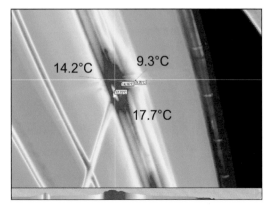

Fig. 7-9 Uninsulated cantilever steel beams traverse the facade; elevated temperatures clearly exhibit the impact of the thermal bridge at the transition.

In comparison to the uninsulated steel beam shown in Fig. 7-9, the thermogram in Fig. 7-11 illustrates a thermally isolated steel beam. Due to the thermal separation, there is no significant temperature rise of the steel beam traversing the building envelope.

In addition to increased heat loss, mold formation, and the general destruction of the construction, another important topic to be considered is the dripping of accumulated condensation in indoor swimming pools. Of course, it is very unpleasant for visitors of a swimming bath when condensation drips from the building elements' surfaces. Moreover, the accumulation of condensation is oftentimes misinterpreted by laymen as the result of a leakage in the construction. This problem has a high importance especially when it comes to glass domes and indirect lights since they oftentimes have a lower thermal insulation than the opaque construction elements and therefore are more likely to generate condensation. Therefore, these construction elements should be equipped either with a warm air current to hinder the generation of condensation or a gutter to drain the condensate. If there is no way to prevent condensation, there is still the option of heating the thermal bridges, for example, with self-regulating heating bands (see the example in [34]).

Fig. 7-10 Alternative to thermal bridges: the thermal isolation of steel beams (source: Co. Schöck, Baden-Baden, Germany).

Date	Temperature boundary conditions		Additional information
	Exterior temperature	Interior temperature	
18.10.2005 morning	approx. +6°C	approx. +22°C	./.

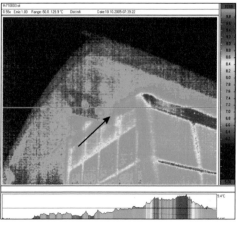

Fig. 7-11 Thermography of a thermally isolated steel beam. The conspicuous area in the thermogram (red coloring) comes from a ventilation grille located under the eaves above the glazing; warm air escapes here.

7.1.4 Solid Round Pillar Traversing the Thermal Insulation Layer

Solid construction elements traversing the thermal insulation layer are oftentimes necessary for constructive or static reasons and therefore unavoidable. A reinforced-concrete pillar traversing the thermal insulation layer is illustrated in Fig. 7-12. In this case the thermal bridge problem shown in the typical warming of the prop-head doesn't lead to constructive defects but is rather relevant as an energetic thermal bridge.

In the case on hand – the pillar is located in an unheated buffer zone (an underground parking lot) – possible measures to lower the energetic heat loss would have to weigh the anticipated savings against the material expenses.

Date	Temperature boundary conditions		Additional information
	Exterior temperature	Interior temperature	
10.03.2005 night-time	approx. +1°C	approx. +22°C	Pillar is located in a naturally ventilated underground parking lot underneath a heated office building

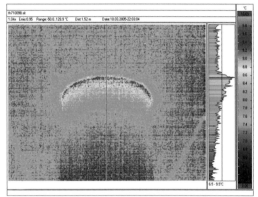

Fig. 7-12 Concrete pillar traversing the thermal insulation layer.

7.1.5 Built-in Roller Blind Box

Within the framework of a survey, the residential building shown below was thermographically evaluated. In order to get a preferably high temperature gradient, the builder-owner set the indoor temperature very high.

In the thermal images, locally elevated temperatures were found especially in the area above the windows. Here the highest temperature was not detected directly at the window but rather slightly above the window. By means of available reference documentation, this phenomenon could be ascribed to the infiltration of warm air from the interior into the roller blind box, which is integrated into the wall construction.

Date	Temperature boundary conditions		Additional information
	Exterior temperature	Interior temperature	
23.01.2005 afternoon	approx. 0°C	approx. +28°C	Exterior shot, relatively high interior temperature due to prior heating

Fig. 7-13 Residential building at the time of the thermography measurement *(left)* and at the time of construction *(right)*: above the windows, insulated roller blind boxes and the circular ring beams (U-shaped shells with reinforced concrete, see **Fig. 7-20**) are located.

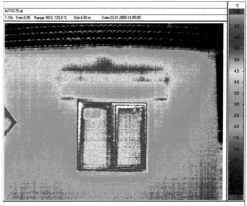

Fig. 7-14 Thermal images: *(left)* overall view; *(right)* windows with visible contours of the roller blind box.

Here, the warm air heated the ring beam above the roller blind box so a local increase of the temperature compared to the undisturbed wall construction of about 2 Kelvin could be found.

Thus, it is very important to understand the general construction of the wall as well as its details for the specific evaluation of a question.

Further analysis has been conducted on the same building at a triangular protruding extension of the building. As illustrated in Fig. 7-15, the building was under direct solar radiation when the thermography measurement was conducted. As already explained in Chapter 6.1.1, outdoor thermography is not meaningful under such circumstances. The thermal image of the wall-to-ceiling crossover was therefore taken from the inside and is shown in Fig. 7-16.

Date	Temperature boundary conditions		Additional Information
	Exterior temperature	Interior temperature	
23.01.2005 afternoon	approx. 0°C	approx. +28°C	Indoor shot. Liquid applied wallpaper made out of viscose, cellulose, and cotton fibers; relatively high air temperature due to prior heating

Fig. 7-15 Triangular annex of the building; (*right*) interior view of the protruding corner region.

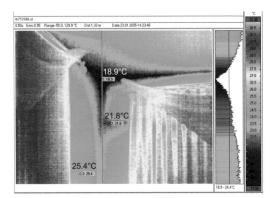

Fig. 7-16 Thermal image showing a significant decrease in the surface temperature distribution directly in the corner.

Fig. 7-17 Photo of the examined wall region taken during the construction process.

On one hand the thermal image exhibits a significant drop in the interior surface temperature in the region of the wall-to-ceiling crossover and corner; on the other hand, it shows a drop in the surface temperature at the bearing point of the lintel/roller blind box.

The decrease in the surface temperature at the ceiling could be explained by a lack of thermal insulation, which wasn't installed over the whole area of the solid ceiling (see Fig. 7-15). Thus, cold outdoor air could undercut the existing thermal insulation and effectuate a cooling of the ceiling.

Apparently, materials with a high thermal conductivity were used for the bearing points of the roller blind boxes (mortar bearing points for adjustment), which led to local material-related thermal bridges. However, those regions represent locally limited weak spots which in general do not cause any damage and can therefore be regarded as noncritical.

7.1.6   Mold Formation at an Exterior Wall Corner: Extensive Evaluation

In a bedroom of a residential building, the renter complained about beginning mold formation. In addition to an evaluation of the heating and aeration behavior and analysis of the building construction, a thermal image was taken during the cold season in the frame of an expert opinion. The evaluation of the thermal image exhibits a conspicuously continuous and uniform cooling in the corner region between the ceiling and the exterior wall. Such uniform temperature differences between the undisturbed wall and its connection to the ceiling can be taken to indicate a constructive anomaly. Together with the builder-owner, the

Date	Temperature boundary conditions		Additional information
	Exterior temperature	Interior temperature	
16.01.2005 midday	approx. 0°C	approx. +22°C	Indoor thermography, exterior wall made of masonry without additional thermal insulation. Low outdoor temperatures were constant for two days before the thermography measurement.

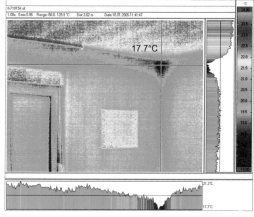

Fig. 7-18 Corner region of an exterior bedroom wall.

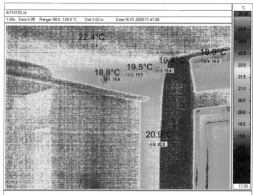

Fig. 7-19 Beginning mold formation in the corner region.

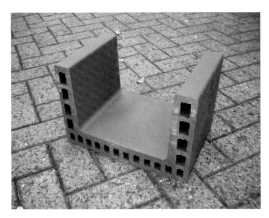

Fig. 7-20 Available shapes of U-shaped shells; (*right*) shell with factory-made thermal insulation.

structure of the construction in the critical area was examined more closely. Evaluating the construction plans revealed that in the exterior section, a continuous ring beam had been arranged on top of the Porotherm blocks (known as "Poroton masonry" in German-speaking countries) inside a U-shaped shell. However, the U-shaped shell had been completely filled with concrete and had not been additionally insulated on the outside, as had been planned.

By means of appropriate software (see [33]), the region of the thermal bridge has been numerically analyzed according to the regulations of DIN 4108-2 (see Chapter 6.2.5).

The existing construction at the crossing between the wall and ceiling is illustrated in Fig. 7-21. Fig. 7-22 shows the computed temperature field with the point of critical surface temperatures.

The minimal surface temperature θ_{si} was calculated to be 11.6°C and thus, smaller than the critical surface temperature $\theta_{si,cr}$ of 12.6°C.

The calculation of the temperature factor f_{Rsi} results:

$$f_{Rsi} = \frac{\theta_{si} - \theta_e}{\theta_i - \theta_e} = \frac{11.6\,°C - (-5\,°C)}{20\,°C - (-5\,°C)} = 0.66 \leq 0.70 \tag{36}$$

Hence, even with proper heating and aeration behavior, the accumulation of condensate and formation of mold can't be completely eliminated. In this case the cause of the mold formation is the ring beam region, which had not been thermally insulated during the construction process.

In comparison to the case above, the temperature field and the critical surface temperatures of the planned construction have also been numerically computed (Figs. 7-23 and 7-24).

The minimal surface temperature θ_{si} is 13.3°C and thus, higher than the critical surface temperature $\theta_{si,cr}$ of 12.6°C, which is not to be undercut. The temperature factor according to DIN 4108-2 [12] is also calculated to an uncritical value ≥ 0.7.

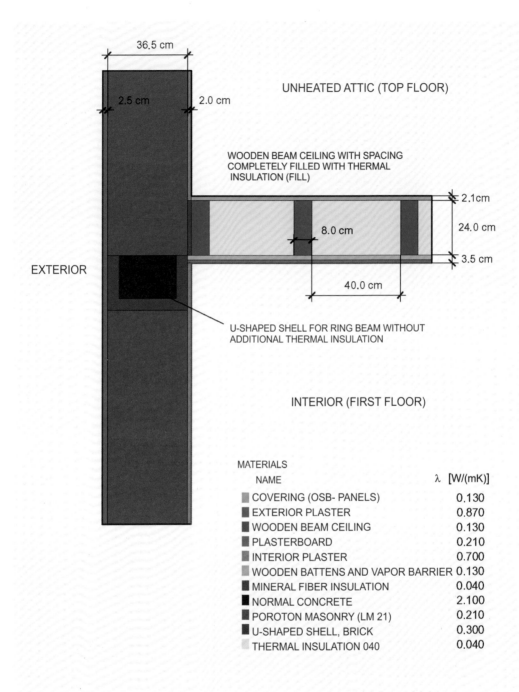

36.5 cm

2.5 cm　2.0 cm

UNHEATED ATTIC (TOP FLOOR)

WOODEN BEAM CEILING WITH SPACING
COMPLETELY FILLED WITH THERMAL
INSULATION (FILL)

2.1 cm

24.0 cm

8.0 cm

3.5 cm

EXTERIOR

40.0 cm

U-SHAPED SHELL FOR RING BEAM WITHOUT
ADDITIONAL THERMAL INSULATION

INTERIOR (FIRST FLOOR)

MATERIALS

NAME	λ [W/(mK)]
COVERING (OSB- PANELS)	0.130
EXTERIOR PLASTER	0.870
WOODEN BEAM CEILING	0.130
PLASTERBOARD	0.210
INTERIOR PLASTER	0.700
WOODEN BATTENS AND VAPOR BARRIER	0.130
MINERAL FIBER INSULATION	0.040
NORMAL CONCRETE	2.100
POROTON MASONRY (LM 21)	0.210
U-SHAPED SHELL, BRICK	0.300
THERMAL INSULATION 040	0.040

Fig. 7-21 Construction without additionally insulated U-shaped shell (existing structure).

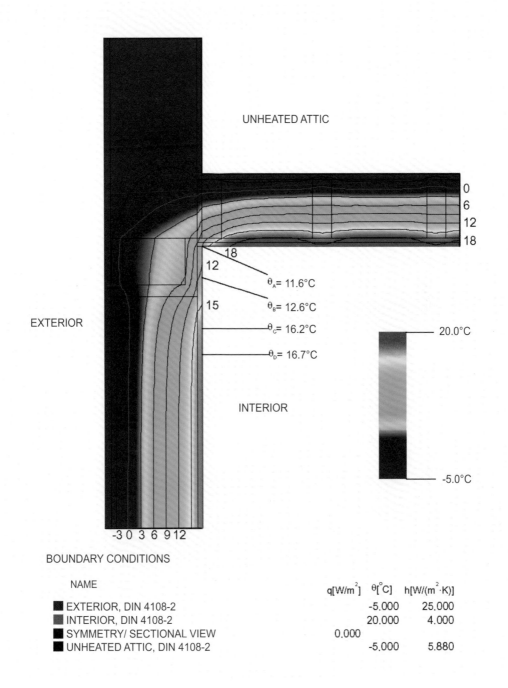

UNHEATED ATTIC

0
6
12
18

18
12

$\theta_A = 11.6°C$

15

$\theta_B = 12.6°C$

$\theta_C = 16.2°C$

$\theta_D = 16.7°C$

EXTERIOR

INTERIOR

20.0°C

-5.0°C

-3 0 3 6 9 12

BOUNDARY CONDITIONS

NAME

	q[W/m²]	θ[°C]	h[W/(m²·K)]
■ EXTERIOR, DIN 4108-2		-5.000	25.000
■ INTERIOR, DIN 4108-2		20.000	4.000
■ SYMMETRY/ SECTIONAL VIEW	0.000		
■ UNHEATED ATTIC, DIN 4108-2		-5.000	5.880

Fig. 7-22 Temperature distribution of the construction without additionally insulated U-shaped shell (existing structure).

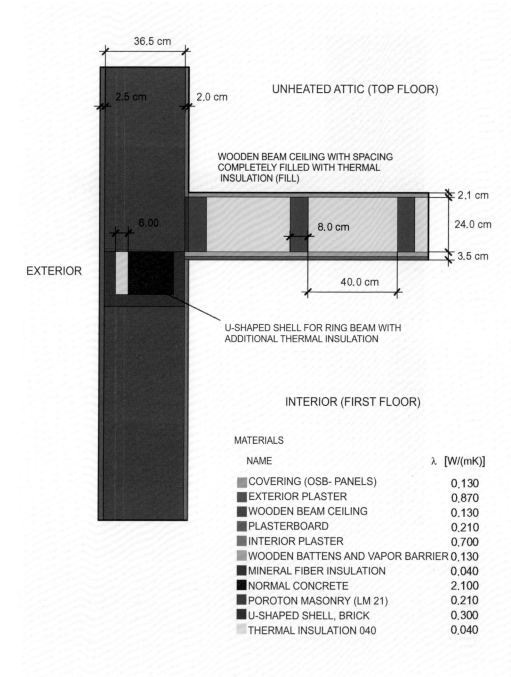

Fig. 7-23 Construction with thermally insulated U-shaped shell (planned structure).

36.5 cm

2.5 cm   2.0 cm

UNHEATED ATTIC (TOP FLOOR)

WOODEN BEAM CEILING WITH SPACING
COMPLETELY FILLED WITH THERMAL
INSULATION (FILL)

2.1 cm

6.00

8.0 cm

24.0 cm

3.5 cm

EXTERIOR

40.0 cm

U-SHAPED SHELL FOR RING BEAM WITH
ADDITIONAL THERMAL INSULATION

INTERIOR (FIRST FLOOR)

MATERIALS

NAME	λ [W/(mK)]
COVERING (OSB- PANELS)	0.130
EXTERIOR PLASTER	0.870
WOODEN BEAM CEILING	0.130
PLASTERBOARD	0.210
INTERIOR PLASTER	0.700
WOODEN BATTENS AND VAPOR BARRIER	0.130
MINERAL FIBER INSULATION	0.040
NORMAL CONCRETE	2.100
POROTON MASONRY (LM 21)	0.210
U-SHAPED SHELL, BRICK	0.300
THERMAL INSULATION 040	0.040

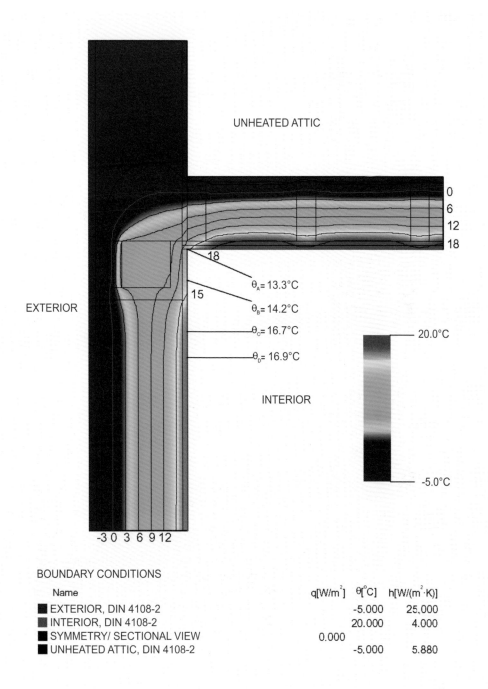

UNHEATED ATTIC

EXTERIOR

INTERIOR

θ_A= 13.3°C
θ_B= 14.2°C
θ_C= 16.7°C
θ_D= 16.9°C

20.0°C

-5.0°C

BOUNDARY CONDITIONS

Name	q[W/m^2]	θ[°C]	h[W/(m^2·K)]
EXTERIOR, DIN 4108-2		-5.000	25.000
INTERIOR, DIN 4108-2		20.000	4.000
SYMMETRY/ SECTIONAL VIEW	0.000		
UNHEATED ATTIC, DIN 4108-2		-5.000	5.880

Fig. 7-24 Temperature distribution of the construction with thermally insulated U-shaped shell (planned structure).

$$f_{R_{si}} = \frac{\theta_{si} - \theta_e}{\theta_i - \theta_e} = \frac{13.3\,^{\circ}\mathrm{C} - (-5\,^{\circ}\mathrm{C})}{20\,^{\circ}\mathrm{C} - (-5\,^{\circ}\mathrm{C})} = 0.73 \geq 0.70 \tag{37}$$

In addition, within the framework of the examination, user behavior was also plotted continuously over an appropriate period of time. For this purpose, data loggers which automatically and in a predetermined interval recorded the values of air temperature and relative humidity were used. The evaluation of this data is illustrated in Fig. 7-25. During the recording period of about three weeks, the average room air temperature θ_i was about 18.6°C and the average relative humidity ϕ_i was about 50.1%. The recorded data shows a significant drop in air temperature and humidity at frequent intervals, which is a typical indication of morning airing. During the day the relative humidity remained fairly constant. During the night, the relative humidity in the bedroom rises to a climax until it drops again due to the morning airing. Altogether, user behavior can be rated as noncritical and does not provoke mold formation.

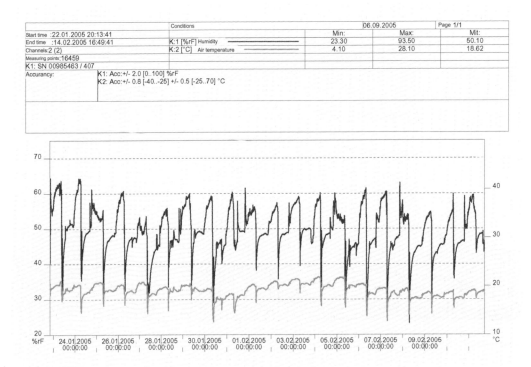

Fig. 7-25 Temperatures and relative humidity during the measurement period, recorded via data logger.

7.2 Use of Thermography in Localizing Air Leaks

7.2.1 Support of Thermography by Means of Pressure Difference Methods (Blower Door Test)

Timber-Framed Buildings: Example no. 1

At a timber-framed residence building with story-high construction parts, the destruction of the wood by fungal decay had been detected. After preliminary inspections, additional thermal images in combination with a blower door test were taken.

First, the thermal image evaluation was conducted under normal conditions, meaning without implementing any excess or low pressure.

Subsequently, a blower door test with excess pressure was conducted, causing leakage of warm interior air through surface defects. In order to achieve the highest possible temperature gradient during the additional thermography analysis, the builder-owner had turned up the heaters to the highest level the day before. At potential surface defects, the warm inside air goes through the construction and escapes outside. At this point, the construction grows warm, which can be visualized by means of thermography. This procedure is illustrated in Figs. 7-28 and 7-30.

Fig. 7-26 Residence building with exposed examination point in the region of the ceiling slab. The external construction is removed all the way to the thermal insulation.

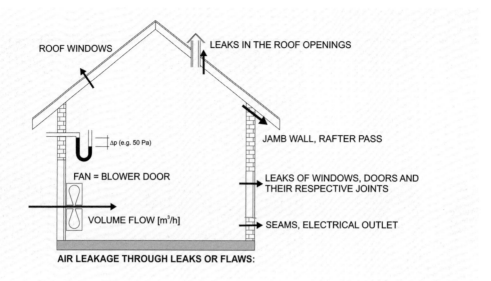

ROOF WINDOWS

LEAKS IN THE ROOF OPENINGS

Δp (e.g. 50 Pa)

FAN = BLOWER DOOR

VOLUME FLOW [m³/h]

JAMB WALL, RAFTER PASS

LEAKS OF WINDOWS, DOORS AND THEIR RESPECTIVE JOINTS

SEAMS, ELECTRICAL OUTLET

AIR LEAKAGE THROUGH LEAKS OR FLAWS:

Fig. 7-27 Principle of the differential pressure measurement (colloquially blower door measurement); here, overpressure measurement.

Recording date	Temperature conditions		Additional information
	Exterior temperature	Interior temperature	
12.11.2004 early morning	approx. −1°C	approx. +26°C	Ideal climate conditions for outdoor thermography (long period of cold and overcast skies), relatively high interior temperature due to intensive heating

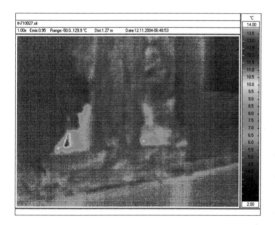

Fig. 7-28 Examination point at zero state (no excess pressure or low pressure).

Afterward, the blower door test was used in the low pressure mode. In this case the pressure inside the building is lower than on the outside. Cold external air can infiltrate into the building through surface defects. In the documented example, this phenomenon became

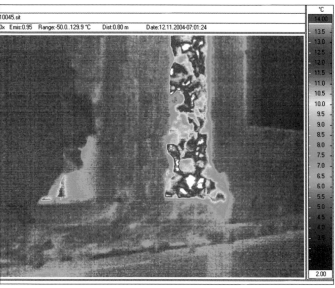

Fig. 7-29 Blower door in position (excess pressure measurement).

Fig. 7-30 Examination point during the blower door test (implementing excess pressure); the right side of the construction part shows a significant temperature rise compared to Fig. 7-28.

visible by a rapid cooling-down of the warmed-up regions (Fig. 7-31). On the interior, the infiltrated external air leads to a cooling of the building part's surface and can be detected by thermography images taken indoors.

As a result of this examination, the significant change in the surface temperature was only existent in the right wooden post-and-beam component (corner component, looked at from outdoors).

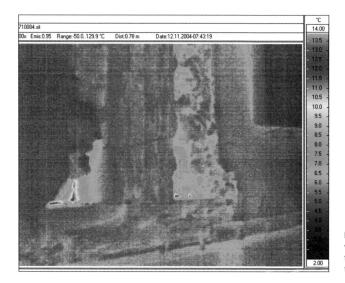

Fig. 7-31 The examination point shows a very rapid cooling-down of the surface temperature after inducing a lower pressure through the blower door test.

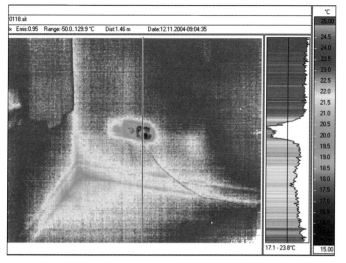

Fig. 7-32 Indoor thermogram of the corner region after inducing low pressure.

The left side of the component (attached parapet component) exhibited nearly identical behavior concerning the surface temperature in all three examinations. Thereby the cause of damage could be narrowed down to the corner component. Further investigation showed that the internally installed vapor barrier had been severely perforated by subsequently installed electrical sockets in the corner region (see Fig. 7-32). Thus warm interior air could escape particularly through the corner region and condense inside the construction, which finally led to the damage of the wooden construction.

With the assistance of the differential pressure method (blower door test) thermography is able to clearly point out existing air leaks and hence to reveal surface defects faster and more easily. When implementing the pressure difference, values higher than 50 Pa should not be exceeded since higher pressure differences might lead to the destruction of the wind proofing or the vapor barrier.

Timber-Framed Buildings: Example no. 2

Thermal image evaluation in combination with a blower door test was conducted on a timber-framed building. For this purpose, the construction was uncovered in the area where the ceiling connects to the wall without destroying the wind proofing or the vapor barrier. Then, with the aid of indoor thermography, the surface temperatures of the region were examined before and during the application of low pressure.

Fig. 7-33 Photograph of the examination area later uncovered within the framework of the thermography analysis.

Date	Temperature boundary conditions		Additional information
	Exterior temperature	Interior temperature	
13.12.2004 evening	approx. −1°C	approx. +25°C	./.

Fig. 7-34 Uncovered examination area between the wall and ceiling and respective thermograms before and during the application of low pressure (about 50 Pa).

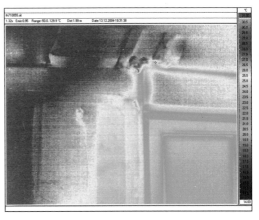

Normal state before the application of low pressure (A).

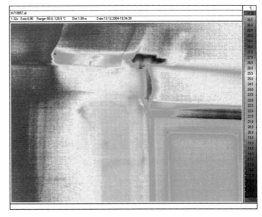

After 3 minutes of applying low pressure (B).

The evaluation of the thermal images shown in Fig. 7-34 can be simplified by means of modern IR systems by applying the difference-image technique. Therefore, a temperature distribution is assumed as the subtrahend and is subtracted in the following thermal images as the minuend. The result is the difference between both temperature distributions.

In order not to irritate the subjective perception of the observer, "cold" colors should be chosen in the case above to visualize the cooling of the surface. "Red" colors could easily lead to the assumption that the surface warms up, even though this is not the case. Thus Figs. 7-35 and 7-36 show the same difference image ($\Delta\theta$ in [K]) with different temperature ranges:

Temperature range: 0°C to −10°C:

The red color in the thermal image in Fig. 7-35 may create the false impression of a warming.

Temperature range: +5°C to −10°C

The thermal image in Fig. 7-36 exhibits fewer red colors. Therefore, the image is perceived to be "colder."

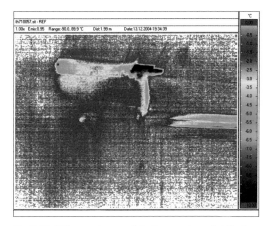

Fig. 7-35 Difference image of the temperature distribution before and after the application of low pressure; calculation rule: A − B = difference image, $\Delta\theta$ in [K].

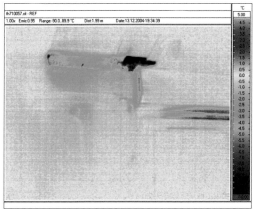

Fig. 7-36 Difference image of the temperature distribution before and after the application of low pressure; calculation rule: A − B = difference image, $\Delta\theta$ in [K].

7.2.2 Air Leaks in the Roof and Other Constructions

Air Leaks in the Roof Area

Roof constructions can also be examined for leakage areas without support to achieve a pressure difference. In most cases, such areas can also be visually detected in the cold season by color changes in areas otherwise covered with frost or snow melting.

Date	Temperature boundary conditions		Additional information
	Exterior temperature	Interior temperature	
25.11.2016 evening	approx. 0°C	approx. +22°C	The photograph was taken 2 days after the thermogram. The defrosted traces on the roof clearly show the areas with heat leakage.

Fig. 7-37 Documentation of heat leakages on a fully heated roof.

Industrial Hall Construction with Different Temperature Ranges and Non-Airtight Ceiling Construction

In an air-conditioned warehouse for foodstuffs (permanently +5°C) with a normally heated office area with ventilation system enclosed in the building (approx. +22°C), condensation was detected and criticized at the transition of the areas of use at the partition wall. In a

Date	Temperature boundary conditions		Additional information
	Exterior temperature	Interior temperature	
20.11.2013	approx. +5°C	approx. +22°C	Thermogram recorded inside a cold store with a view of the partition wall to the normally heated office area.

 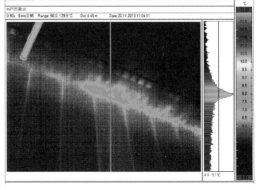

Fig. 7-38 View of the partition wall between the office area (22°C) and the air-conditioned warehouse (+5°C); selective heating in the upper corner area wall/roof.

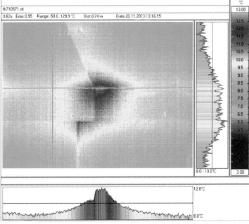

Fig. 7-39 Detail transition wall/ceiling: warm air from the (warm) office area reaches the air-conditioned (cooler) area and condenses there; water run-off traces are the consequence.

first guess as to the cause of the moisture events, leaks in the waterproofing in the roof area were suspected. With the help of thermography, it was proven that the internal partition wall between the office/storage area showed air leaks, which were indicated by conspicuous hot spots in the partition wall area.

The reason for these phenomena was a non-airtight connection of the continuous steel trapezoidal roof between the different climate zones. The warm office air reaches the year-round air-conditioned +5°C area through a slight overpressure of the ventilation system and condenses there. This leads to the detected water run-off tracks.

Fig. 7-40 Filling of the bead in the trapezoidal steel sheet without airtight seal, measurement of the natural air emission at the measuring point (here v = 0.20 m/s).

Non-airtight Connections of Brick Linings in Timber-Framed Buildings

Timber-framed buildings are characterized by the fact that fillings are introduced into the load-bearing wooden structure, which can consist of plaster or brick linings, for example. The existing joint in the area of the material change represents a challenge with regard to airtightness. In such buildings, there are very often areas where convective heat transfer takes place. With the aid of thermography, such areas can be well documented.

Practical Examples

Date	Temperature boundary conditions		Additional information
	Exterior temperature	Interior temperature	
09.11.2016	approx. +2°C	approx. +22°C	

Fig. 7-41 brick lining on a timber-framed wall with documented local air leakage.

7.2.3 Air Leaks along Windows and Doors

Air leaks along windows and doors lead to elevated ventilation heat loss and air drafts. Along outer windows and doors, air leaks can cause significant degradation of the soundproofing properties. By means of thermography, leaks can be quickly detected and documented for specific repair measures.

Date	Temperature boundary conditions		Additional Information
	Exterior temperature	Interior temperature	
27.01.2005 evening	approx. 0°C	approx. +23°C	Underfloor heating in this area

Fig. 7-42 Entrance door with air leak along door rabbet. The horizontal blue area on the right side of the door shows the mail slot.

7.3 Use of Thermography to Detect Permanent Moisture Penetration

Thermal conductivity λ [W/(mK)] of construction materials is generally given for materials in dry conditions. Dry materials achieve their thermal insulating effect through very small air particles that are captured within the mineral matrix.

When materials get moist, the air particles and pores in the mineral matrix partially fill up with water, which leads to an increasing thermal conductivity λ. Thus, moist materials conduct and spread heat better than the dry materials. Due to the higher water content and the related higher heat storage capacity, the thermal inertia of the materials rises. Furthermore, water can evaporate at the surface of the moist material where the needed heat is taken from the material and hence cools it (principle of cooling by sweating). This heat loss caused by phase transition is significant in more precise calculation methods such as discourses about thermal-hygric transportation and is covered extensively in [29] and others.

With the aid of thermography, temperature differences caused by moisture penetration can be detected at the surface, and the dimension of moisture penetration can be described. However, the moisture content cannot be quantified by this technique and thus has to be determined using other techniques.

7.3.1 Defects Due to Moisture Penetration at a Basement Wall

Example no. 1

In the privately used basement of an old building, a survey was conducted before the planning of a renovation in order to document its state. It had to be determined whether the basement walls had a vertical as well as a horizontal sealing and if so, what their condition was. In order to get an overview of the existing moisture penetration and to localize particularly critical wall areas, thermography was used in addition to other methods.

In combination with a series of data from a moisture measuring device, thermal images showed significant moisture penetration concentrates on the floor area, as expected due to stains visible in photographs taken at the time. Thus, by means of thermography, one could get a quick impression of the building fabric.

Date	Temperature conditions		Additional information
	Exterior temperature	Interior temperature	
29.06.2005 evening	approx. +23°C	approx. +19°C	Privately used basement room of an existing building; additional measurements were carried out by means of a dielectric moisture measuring device.

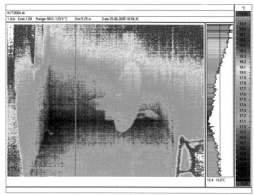

Fig. 7-43 Damp areas of a basement exterior wall; the shape of the moisture horizon is already visually detectable.

Example no. 2

Damage due to humidity was detected on the basement walls of a residential building constructed in 1932. In order to get a quick overview of the severe damp areas, thermography technique was applied.

As shown in Fig. 7-44, the severely damp region stands out with very low surface temperatures in the thermal image. In addition to the thermal images, the relative moisture distribution of the walls was measured by means of a dielectric measuring device. The measurement of this device is based on the principle of the capacitive electrical field. The measuring field develops between a ball electrode of the device and the mass of the examined area. Fluctuations in the electrical field due to a change of material and/or moisture is digitally displayed in a range between 0 and 199 digits. This is a relative measuring, which means that the displayed value is not the absolute moisture content in g/cm³ but only a value indicating the dimension of the construction element's moisture content considering the gross density of the material.

However, in general it can be said that the measured value of the measuring device rises with higher moisture content and gross density of the material, whereby possibly existing contamination with salts (changing of conductivity) influences the measuring value.

It could be seen that the measuring values of the dielectric measuring device obtained at different heights of the basement walls coincided to a relatively high degree with the colder surface regions captured by means of the thermal imaging technique.

Date	Temperature boundary conditions		Additional information
	Exterior temperature	Interior temperature	
23.04.2005 evening	approx. +18°C	approx. +16°C	./.

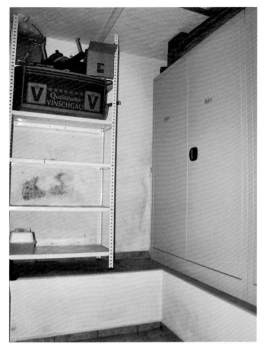
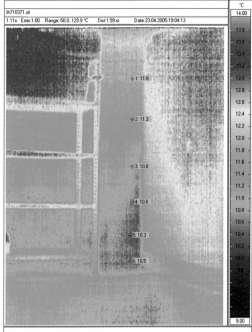

Fig. 7-44 Stains on the plaster show the moisture influence.

Fig. 7-45 Measuring the moisture, in the lower wall region about 134 digits, in the higher (dryer) wall region about 107 digits.

7.3.2 Defects Caused by Moisture Penetration on Flat Roofs

With the help of thermography, it is possible to make areas of moisture penetration visible. As a classic example, examinations of flat roof are often used to find leaks (see Fig. 7-46).

In addition to the increased thermal conductivity (material thermal bridge), the basis for the visibility of conspicuous areas is also the different heating and cooling behavior of the moist areas as a result of the slower (transient) temperature behavior.

Fig. 7-46 Documentation of a leakage on a flat roof (source: with kind support of FLIR Systems GmbH, www.flir.com).

As a result of solar radiation during the day, humid areas (or water inclusions) warm up more slowly due to their increased heat capacity; at night, they maintain the temperature longer during cooling (see also thermal diffusivity a, see for example, the *Bauphysik-Kalender 2005* [35]).

The basic principle can be clearly demonstrated with a simple experimental setup. In a polystyrene plate, cavities of different depths, filled with air or water, were introduced. The cavities were covered with a transparent foil to simulate a seal (see Fig. 7-47).

After filling the examination sites with water, a local heating of the examination sites was carried out with the help of a fan heater. This represents the heating of the roof by the sun's radiation during the day. Due to the inertia and the higher mass of the area with the water confinement that can be stored, these areas do not heat up so strongly (see Fig. 7-49 left). After approx. 10 minutes had passed, the thermogram shown in Fig. 7-49 right was recorded. In the areas without water inclusion, the temperature increase is reduced, while the water-filled areas are still very recognizable as warmer areas. The thermography of (leaky) flat roofs can therefore produce thermograms in which the moisture points appear as colder

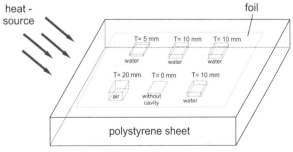

Fig. 7-47 Polystyrene plate with covered examination areas: air- and water-filled cavities.

Date	Temperature boundary conditions		Additional information
	Exterior temperature	Interior temperature	
08.05.2006 midday	./.	approx. +20°C	Interior view of the test specimen, First: filling water in the cavities, Second: heating the test area by using a fan heater

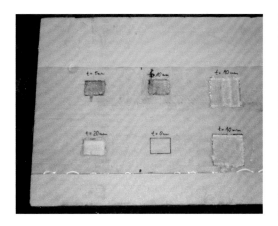 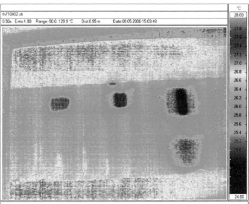

Fig. 7-48 Thermogram after filling up with water: only the water-filled cavities (imitation of the places where water is under the seal) show up as cooler areas.

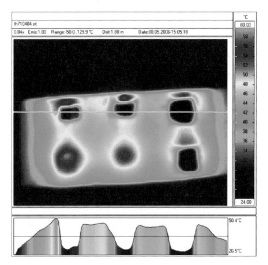 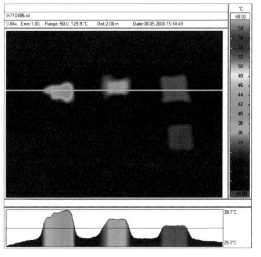

Fig. 7-49 Thermogram shortly after heating of the test sites (*left*); thermogram recorded 10 minutes later (*right*): the water-filled areas are warmer than the previously very warm areas without water inclusion.

areas after the beginning of solar radiation and general warming of the roof or as warmer areas in the early evening after the sun has set and the surface has cooled down. In addition, it should be mentioned that with the help of thermography, leak detection is sometimes not possible; other promising methods exist here (see *Bauphysik-Kalender 2002* [50]).

During the practical leakage search on flat roofs, the user is confronted with the problem that suitable, higher adjacent areas are required for the roof area to be analyzed. Such a structural situation is usually limited or nonexistent. Recent advances in remote-controlled drones have opened up new possibilities for thermography when such technical aids are used.

With the help of a drone equipped with an infrared camera (Flir Tau2, uncooled VOx [vanadium oxide] microbolometer core, 640 × 512 pixel resolution, 50 mK, weight 45 g) from the Institute for Meteorology and Climatology of the Leibniz Universität Hannover, a flat roof inspection test was carried out.

Fig. 7-50 Drone system (manufacturer DJI) with thermal camera Flir Tau2 (company Teax).

Date	Temperature boundary conditions		Additional information
	Exterior temperature	Interior temperature	
22.11.2016 midday	approx. +12°C	approx. +22°C	test flight for the principal possibility of a flat roof inspection

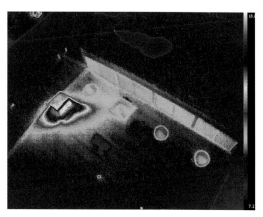

Fig. 7-51 Thermogram with the aid of a remote-controlled drone: hot spot in the area of an exhaust ventilation system; (*top left*) the drone can be seen in flight.

Fig. 7-52 Thermogram generation with the help of a remote-controlled drone (clear picture was taken from the ground view on another day).

This drone was flown by remote control; the thermogram transmitted live to a monitor. It was determined that at least two people must be present during such a thermography: one to operate the drone, and a second to interpret the live image and give feedback to the operator on the drone's position and any instructions to redirect the drone.

Fig. 7-53 Live image transmission from the drone to the monitor (*left*); view of the drone (*right*).

Overall, it is to be expected that an airborne thermography system has great potential for many possible applications. In this area, significant improvements in operation and performance are expected over time.

7.4   Localization of Construction Details by Means of Thermography

7.4.1   Documentation of the Pipe Run of Underfloor Heating

For subsequent installation of convection heaters, the pipe run of existing underfloor heating had to be determined. Furthermore, the builder-owner wanted to know in which part of the floor the warmer flow of the heating pipes had been installed.

Thermography is a very good way to visualize the pipe run of underfloor heating. In order to get the highest possible temperature gradient between the heating pipes and the rest of the floor, the thermography measurement should be carried out in the heat-up phase. In the case considered here, the measurement was carried out on a wood parquet floor. About two days before the measurement the heat was turned down as far as possible in order to cool down the floor construction. On the day of the measurement, the flow temperature of the underfloor heating was set to the maximum. Thus, the pipe run and the warmer flow could be localized by means of the thermography images.

In addition, it should be mentioned that the pipes attached to the heating distributor can also easily be checked (Fig. 7-56).

For the installation of a fireplace within a living room with underfloor heating, a load-bearing construction of corings from the fireplace down to the ground slab had to be drilled. In order not to damage the underfloor heating, the pipe run had to be known in advance. With the aid of thermograms, the flow and return pipes were localized and the layout of the pipe run was documented on the floor with glue dots (Fig. 7-57). Due to the heating process, flow pipes

Date	Temperature boundary conditions		Additional information
	Exterior temperature	**Interior temperature**	
09.12.2003 midday	approx. −4°C	approx. +20°C	Indoor thermography of underfloor heating, beginning of the heating with maximum flow temperature on the day of on-site inspection

 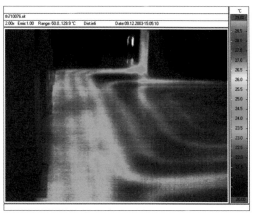

Fig. 7-54 The characteristic geometry of the underfloor heating pipe run becomes visible in the thermogram.

Fig. 7-55 The location of the warmer flow is clearly visible on the right side of the hallway.

Date	Temperature boundary conditions		Further information
	Exterior temperature	Interior temperature	
27.01.2005 evening	approx. −1°C	approx. +20°C	Heating distributor of a residential building

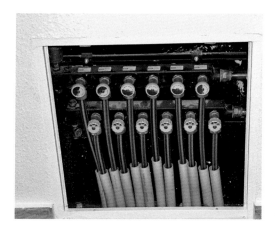
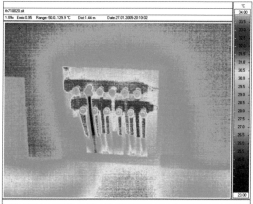

Fig. 7-56 An underfloor heating distributor; one valve of the pipes (*second from left*) is unopened.

become visible quickly and clearly; return pipes, which are placed between the flow pipes, become dimly visible only after a delay (Fig. 7-58). As a result, it was decided to place the corings relatively close to the highly visible flow pipes. The corings were made without any damage, and the fireplace was installed (Fig. 7-59).

Date	Temperature boundary conditions		Additional information
	Exterior temperature	Interior temperature	
02.09.2007 afternoon	./.	approx. +23°C	Indoor thermography of underfloor heating pipes, beginning of the heating with maximum flow temperature on the day of the on-site inspection

Fig. 7-57 Pipe run shown under the parquet floor (*right*). The location of the heating pipes is documented with the aid of glue dots on the floor (*left*).

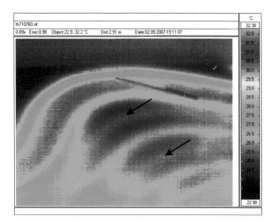

Fig. 7 58 The return pipes are more difficult to detect because they only become dimly visible after a delay. Tip: Drill holes should be located relatively close to the highly visible flow pipes.

Fig. 7-59 Arrangement of the coring to transfer the weight of the fireplace to the ground slab (*left*); first heating of the installed fireplace (*right*).

7.4.2 Exterior Walls with Heating Pipes Installed Beneath the Plaster

Especially when it comes to planning the renovation of old buildings, it is necessary to document the current state of the installed building technical equipment and appliances. In many cases, heating pipes installed beneath the plaster of exterior walls can easily be detected and documented by means of thermography.

Date	Temperature boundary conditions		Additional information
	Exterior temperature	Interior temperature	
27.01.2004 early morning	approx. −5°C	approx. +22°C	./.

Fig. 7-60 The thermogram clearly shows the supply pipes of the heating system on the wall surface.

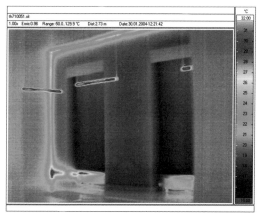

Fig. 7-61 Also in this example, the supply pipes of the heating system can be clearly seen on the wall surface.

7.4.3 Load-Bearing Anchors in Three-Layer Elements

On a residential high-rise building constructed with large panel elements, the location of load-bearing anchors and mounting elements holding the three-layer elements had to be determined. The facade construction is made of sandwiched elements consisting of a load-bearing layer, a weatherproofing layer, and a thermal insulation layer in between. The weatherproofing layer on the outside is mounted with the aid of load-bearing wall plugs and torque anchors which in the thermogram are recognizable as point thermal bridges.

For examinations like these, thermography offers a good nondestructive, comfortable, and efficient way to locate concealed construction parts. In the present case, the weather boundary conditions during the thermography examination were very good (low temperatures and overcast sky in the early morning). This resulted in clear thermograms, depicting the mounting elements as hot spots, standing out even at a distance.

Date	Temperature boundary conditions		Additional information
	Exterior temperature	Interior temperature	
27.01.2004 early morning	approx. −5°C	approx. +22°C	./.

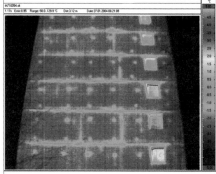

Fig. 7-62 In the thermal image, metallic construction parts are visible through the weatherproof layer as points of elevated surface temperatures.

Fig. 7-63 In addition, thermography showed significant heat loss in the area of the lowest horizontal joint of the construction elements.

7.4.4 Framework Constructions Made of Wood or Concrete

Thermography can be used to examine existing buildings in order to identify the position and structure of frame structures hidden behind plaster or structural components. As an example, Fig. 7-64 shows a residential building with timber frame construction concealed behind plaster on the underside.

The locating of timber frame by means of thermography can also be conducted in summertime, since, due to the different heating and cooling performance of building materials (heat storage capacity), the temperature difference can become visible (see Fig. 7-65).

Date	Temperature boundary conditions		Additional information
	Exterior temperature	Interior temperature	
13.11.2005 evening	approx. +6°C	approx. +22°C	./.

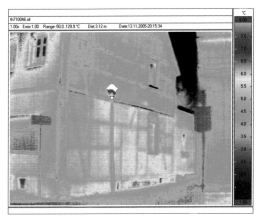

Fig. 7-64 Visualizing concealed timber frame constructions by means of thermography (image taken in November).

Fig. 7-65 The same concealed timber frame construction shown in an image taken in summer.

Practical Examples

Within the framework of existing building surveys, the supporting structure and the thermal bridges that are usually present can be made visible, especially on old, uninsulated buildings. One such task was, for example, the analysis of weak points in building physics for the conversion of the former Tempelhof airfield in Berlin. The listed building was thermographically examined and the conspicuous areas marked in order to be able to suggest possible improvement measures. The existing structure is reinforced concrete clad with shell limestone.

Date	Temperature boundary conditions		Additional information
	Exterior temperature	Interior temperature	
15.03.2013 morning	approx. −4°C	approx. +12°C/+20°C	completely cloudy sky

 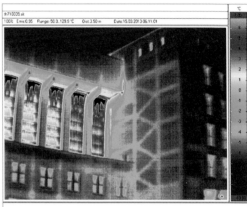

Fig. 7-66 Thermographic visualization of reinforced concrete structures under natural stone slabs.

Fig. 7-67 Reinforced concrete structures under natural stone slabs made visible with the help of thermography; the relatively wide structures of the thermal bridges are conspicuous.

7.4.5 Temperature Distribution of Steel Girders of an Extensively Glazed Open-Sided Floor

For the stability analysis of a steel-framed open-sided floor of a historical building, it had to be determined if the assumptions made about temperature stresses due to thermal layering of the air are plausible. For this purpose, diverse measurements were carried out. In addition, a thermography measurement was carried out at different times in order to determine the temperature differences between the warmer arches at the top and the ties at the bottom of the tie arch construction.

It was shown that the thermal layering of the well-ventilated open-sided floor has led to temperature differences within the construction's load-bearing components.

Date	Temperature boundary conditions		Additional information
	Exterior temperature	**Interior temperature**	
21.05.2003 afternoon	approx. +18°C	approx. +22°C	Overcast sky, only diffuse irradiation

Fig. 7-68 Tied arch construction of an open-sided floor.

Fig. 7-69 Thermal image of the tied arch construction: the temperature of the steel profiles rises with the height of the construction, here about 3 Kelvin.

7.4.6 Load-Bearing Anchor of a Rear-Ventilated Exterior Wall Construction

A rear-ventilated facade construction was installed on an exterior wall. The metal sheet construction, consisting of corrugated sheets and non-bearing pilaster strips, is mounted on a subconstruction which itself is mounted on a massive load-bearing concrete wall by means of point load-bearing anchors.

Within the framework of planning to open the construction, the location of the anchors going through the thermal insulation had to be determined. The thermography was conducted from inside, since in general, rear-ventilated facades exhibit a homogenization of the "cold"

Practical Examples

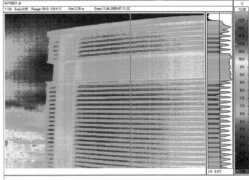

Fig. 7-70 View onto the rear-ventilated facade construction made of corrugated sheets. The angle of perspective changes continuously between 0° and 90° from one corrugation to the next (also see Fig. 2-14).

Fig. 7-71 Thermal image of the rear-ventilated facade taken outdoors. The temperature distribution is falsified by the reflection and angle dependence of the emissivity; see Fig. 2-15.

exterior surface and thus, thermal bridge effects cannot be visualized by means of outdoor thermography (see the example discussed in Chapter 8.1.3). Moreover, due to reflection phenomena and the angle dependence of the emissivity, it is hardly possible to get plausible results out of a thermogram based on a globally valid emissivity (see Fig. 7-71).

The indoor thermography of the wall (see Fig. 7-72) reveals a checkered distribution of the inside surface temperatures. The blue areas signify cooler regions with higher thermal transition due to the installed point anchors. This example clearly shows the effect of the freely customizable color range, since the effective temperature difference between the blue (cooler) and green (warmer) areas is about $\Delta\theta \approx 0.7$ K.

Date	Temperature boundary conditions		Additional information
	Exterior temperature	Interior temperature	
11.04.2005 early morning	approx. +6°C	approx. +25°C	Relatively high interior temperature due to preceding heating

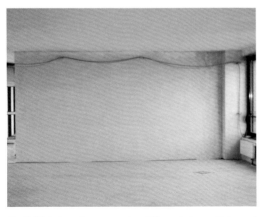

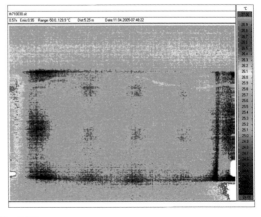

Fig. 7-72 Indoor thermography of the exterior wall shown in Fig. 7-70.

7.4.7   Congestion in Pipes

Congested pipes can lead to sizable damage of the building fabric and the interior installations of the building. Possibly, the location of the congestion can be quickly detected by
means of thermography. In the case below, subsequent cleaning measures showed that
leaves had settled in a pipe and thus had functioned as moisture storage. The decreased
surface temperatures of plastics, which generally exhibit poor heat conduction, could be
made visible with the aid of a thermogram.

Date	Temperature boundary conditions		Additional information
	Exterior temperature	Interior temperature	
29.06.2005 evening	approx. +23°C	approx. +19°C	Basement room under a terrace

 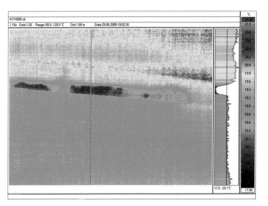

Fig. 7-73 Drain pipe congested with leaves. The pipe drains a gutter installed on the terrace above the basement ceiling.

7.5   Quality Assurance during Construction Works by Means of Thermography

7.5.1   Analysis Before and After an Energy-Focused Building Refurbishment

Within the framework of a refurbishment consulting focusing on building physics and
energy-effectiveness, a building fabric was examined by means of thermography. With the
aid of the thermograms, the builder-owner could be shown that the building fabric (construction year 1932) in no way meets the current legal requirements for thermal insulation.
As an example, Fig. 7-74 shows a thermogram of the existing building fabric. The living room
is on the left side of the building in the perspective. The radiator is located directly under
the window and exhibits a significant local temperature rise of the surface temperature in
the thermal image. During the measurement, the room on the right side of the image hadn't
been heated as planned and thus clearly stands out against the heated room on the left.

There is a bathroom in the expanded area of the attic story (area left of the skylight to about 1 m right of the skylight). In this room, an air leak can be seen in the region of the connection between the ceiling and the attic. This air leak has already led to a significant heating of the roof tiles (see Fig. 7-75). By means of thermography, this leak could be precisely located quickly.

Date	Temperature boundary conditions		Additional information
	Exterior temperature	Interior temperature	
08.03.2003 evening	approx. −3°C	approx. +22°C	Construction year 1932 (hollow blocks made of concrete)

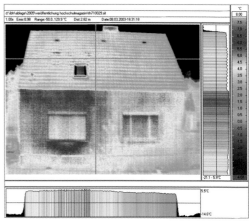

Fig. 7-74 Thermal image before the energy-focused building refurbishment (March 2003).

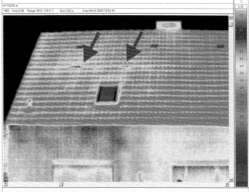

Fig. 7-75 Increased surface temperatures at the connection between the ceiling of the bathroom and the attic (March 2003), reflection of the clear sky in the skylight.

After the refurbishing the thermal insulation of the exterior walls (adding a thermal insulation composite system) and the roof, another thermography measurement was carried out in November 2004 for documentation. Fig. 7-76 shows the thermogram in the same color-temperature scale as Fig. 7-74. The insulated areas exhibit a uniform surface temperature without any detectable voids, which is characteristic for a well-insulated building. At the time of the measurement, the plinth beam hadn't been insulated yet, which is why there is still a significant temperature drop in this area.

A characteristic phenomenon for well-insulated buildings which are free of defects is the accumulation of outside air warmed up by thermal losses under protrusions such as roof overhangs and window lintels. The escape of the warm and hence lighter air can be restrained by the construction form and thus leads to a slight warming of these "sheltered" areas. In thermal images these areas then appear warmer. In this case, this effect doesn't signify a weak spot and has to be considered accordingly, when interpreting thermograms. At this point, it has to be mentioned that this phenomenon only occurs when there is no wind; in the case of higher air ventilation at the construction surface, the screen of warm air cannot develop. However, if there are defects or air leaks in the considered area, the cause of the locally elevated temperatures has to be examined within this context.

After the refurbishment, the building is still heated by a storage heater. Comparing the climate-adjusted annual heating energy consumption of the years 1999–2004 (before refurbishment) to the first year with thermal insulation (2004–2005) revealed an effective saving of the consumed amount of electricity for heating of about 30%. This trend has been confirmed in subsequent years.

Date	Temperature boundary conditions		Additional information
	Exterior temperature	Interior temperature	
13.11.2004 evening	approx. −1°C	approx. +22°C	Plinth beam in the basement not yet insulated; warm area on the roof is caused by a vent pipe from the bathroom in the attic

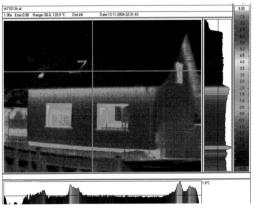

Fig. 7-76 Thermal image after adding thermal insulation onto the roof (20 cm, thermal conductivity 045) and onto the exterior walls (10 cm, thermal conductivity 0.040 W/(mK)). The plinth beam was not insulated at the time of the measurement (November 2004).

In the further course of the refurbishment, the plinth beam was also insulated. The related thermal image is shown in Fig. 7-77. The thermal image is characteristic for plinths: A screen of warm air develops at the undercut between the plinth and the thermal insulation of the wall (see also Chapter 6.2.4). Furthermore, the thermogram shows the location of the mounting bolts for the thermal insulation composite system.

Date	Temperature boundary conditions		Additional information
	Exterior temperature	Interior temperature	
13.11.2005 nighttime	approx. +2°C	approx. +20°C	./.

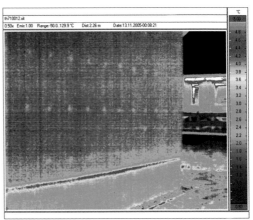

Fig. 7-77 Thermal image after adding thermal insulation to the plinth (6 cm, thermal conductivity 0.040 W/(mK), November 2005).

7.5.2   Inspection of a Building with a Thermal Insulation Composite System

Under the precondition of the right climate boundary conditions, thermography can be a very useful tool for the quality check of thermal insulation measures. In the case of poor workmanship, defects (such as insulation panels not installed flush to one another) show as linear warm spots in the thermal image. The following example of an apartment house showed a single local defect between two windows.

The only conspicuous spot in the thermal image is the clearly warmer basement exterior wall. This area had not been insulated, since a thermal insulation layer installed underneath the basement ceiling defined the boundary of the heated zone. However, after evaluating the thermal image, this decision should be reconsidered in terms of a cost-benefit calculation.

The following images can be used to explain the so-called distance effect emerging in thermography systems: the surface temperature θ_{se} determined from the thermogram depicted in Fig. 7-78 is about 3.6°C. When decreasing the measuring distance, the surface temperature

θ_{se} increases to 4.4°C (Fig. 7-79) and 6.2°C (Fig. 7-80). This is not a measuring error, but is due to the fact that the size of the real measuring area assigned to one pixel of the thermography system's receiver unit increases with the measuring distance. Hence, the integration over the area is finer and more detailed for close-up images (see Fig. 3-11). This is the reason for the different surface temperatures determined in the case below (outdoor thermography of a thermal bridge).

Date	Temperature boundary conditions		Additional information
	Exterior temperature	Interior temperature	
18.02.2005 morning	approx. + 3.5°C	approx. +22°C	./.

 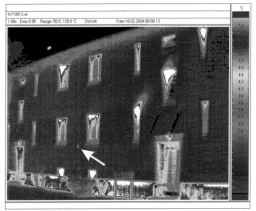

Fig. 7-78 Old building with subsequently installed thermal insulation composite system: the arrow in the thermal image indicates an anomaly within the facade area (between two windows; measured temperature θ_{se} = 3.6°C).

 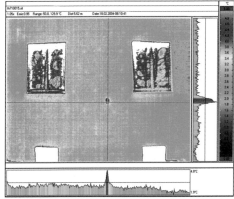

Fig.7-79 More detailed picture of the anomaly detected in Fig. 7-78: the elevated surface temperature θ_{se} between the windows stands out against the rest of the facade (measured temperature: 4.4°C).

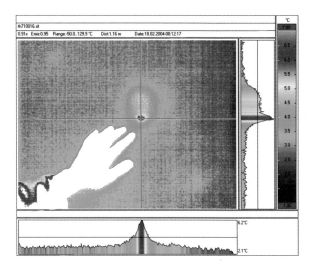

Fig. 7-80 Close-up of the anomaly: in terms of energy efficiency this thermal bridge within the thermal insulation composite system is not critical (measured temperature θ_{se} = 6.2°C, see the peak value in the vertical display). This point thermal bridge is a hole in the insulation layer caused by a bearing anchor of the scaffolding; it was subsequently filled with plaster.

7.5.3 Acoustic Bridges in Parting Lines of Terraced Houses

Measuring the airborne sound reduction indices R'_w in a terraced housing estate revealed poor values of the airborne sound reduction for some of the partition walls between the terraced houses, even though the structure was the same for all partition walls (for the wall structure, see Fig. 7-81). While some building partition walls of the still unoccupied houses had airborne sound reduction indices of R'_w = 66 dB, others only reached insufficient values of R'_w = 54 dB.

Due to the way the building shell was constructed, it could be assumed that when pouring the concrete for the ceilings, some clumps had fallen into the parting spaces between the masonry walls of the terraced houses. When this happens, a significant degradation of the airborne sound reduction index is very likely.

In order to localize acoustic bridges by means of thermography, a significant temperature difference between the two sides of the partition wall is needed. To do so, the apartment on one side was heated to a room temperature of about 32°C ten hours before the measuring. In the apartment on the other side of the partition wall, the room temperature was the same as the exterior temperature during the on-site inspection, which was about 15°C.

The evaluation of the thermal images revealed local hot spots in areas where missing insulation or sound bridges had previously been discovered by endoscopic examination or exposure of the structure (see Fig. 7-82). With the help of the thermographic examination, numerous other critical sound bridges were detected and later removed in this construction project.

Additionally it must be mentioned that the airborne sound reduction index decreases considerably even for small point acoustic bridges. However, due to their marginal thermal bridge effect, they are not visible in thermal images in most cases. Hence, the application of thermography is probably not appropriate for detecting all acoustic bridges.

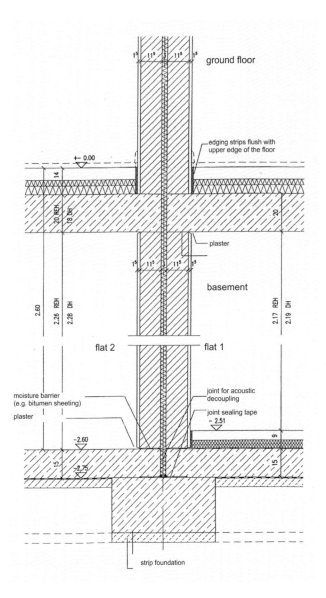

ground floor

edging strips flush with
upper edge of the floor

+ 0.00

plaster

basement

flat 2   flat 1

moisture barrier
(e.g. bitumen sheeting)

plaster

joint for acoustic
decoupling

joint sealing tape
− 2.51

−2.60

−2.75

strip foundation

Fig. 7-81 Planned construction of the
terraced house partition walls.

Date	Temperature boundary conditions		Additional information
	Air temperature in apartment no. 1	Air temperature in apartment no. 2	
22.05.2003 at night	approx. +15°C	approx. +32°C	Generation of a significant temperature difference by setting up a hot air blower in apartment no. 2. The time of measuring (in the month of May) is not optimal, since due to the relatively high exterior temperature, the generation of the temperature gradient was linked with high cost and effort.

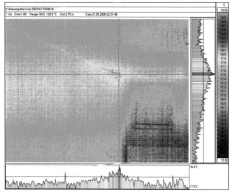

Fig. 7-82 Joint of the basement concrete ceiling and the masonry partition wall, elevated surface temperature near the window: a subsequent uncovering of the construction part revealed missing thermal insulation and clumps of concrete at the height of the ceiling.

7.5.4 Missing Thermal Insulation and Air Leaks in Drywall Construction

In many cases, drywall construction is used for the development of attics. Due to restricted space respecting the roof construction, special attention should be paid to the installation of continuous thermal insulation. Fig. 7-83 shows a typical thermogram of continuous insulation with local defects. The cold spot in the vertical section of the considered evaluation area exhibits a temperature difference of about 6.5 Kelvin compared to the undisturbed area.

After evaluating the results of the thermography measurement, the building owner stuffed additional cellulose insulation material into the detected void. In the following year, this area was again examined (see Fig. 7-84). The area then showed a smooth temperature

Date	Temperature boundary conditions		Additional information
	Exterior temperature	Interior temperature	
27.02.2005 afternoon	approx. −5°C	approx. +25°C	At the time of the on-site inspection, the builder-owner set the room temperature higher than usual.

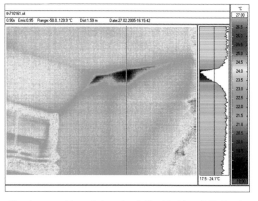

Fig. 7-83 The cold spot in the joint between the wall and the ceiling is caused by cellulose insulation that is not fit closely; the temperature difference is about 6.5 Kelvin.

decrease, which is typical for such edges. When using the same temperature scale as that used in Fig. 7-83, an evaluation of the area is not possible due to the insignificant difference in temperature levels. When shifting to an appropriate temperature scale, it can be seen that there is no longer a significant cold spot. The vertical section at the same spot only indicates a temperature difference of about 2.5 Kelvin.

In order to avoid air drafts, heat loss, and defects within a construction, it is very important to ensure its airtightness. Fig. 7-85 shows a double socket without a draft-proof seal for the cavity wall box.

Date	Temperature boundary condition		Additional information
	Exterior temperature	Interior temperature	
14.01.2006 afternoon	approx. −3°C	approx. +21°C	Unlike the first on-site inspection (25.02.2005) the normal interior temperatures have been set

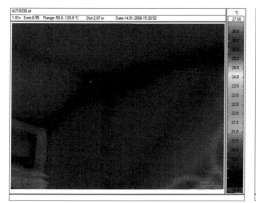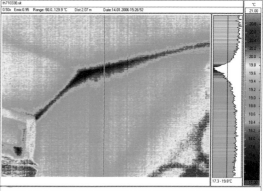

Fig. 7-84 Inspection of the fixed defect shown in Fig. 7-83: (left) for the same temperature range used in Fig. 7-83, it is impossible to interpret the thermal image; (right) using an appropriate temperature scale, the cooling in the edge between wall and the ceiling is smooth; the temperature difference is only about 2.5 Kelvin.

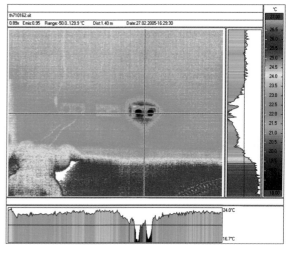

Fig. 7-85 Air leak around an electrical socket.

Practical Examples

7.5.5 Missing Thermal Insulation in a Window Reveal

The renter of an apartment complained about mold formation in the transition between the window frame and the reveal in his bedroom (Fig. 7-86). Within the framework of an evaluator's report, it had to be investigated whether the mold formation had been caused by defects within the construction or by the renter's negligence. Thermography was used for the examination, among other measuring techniques. Both indoor and outdoor thermography were conducted.

The thermal image clearly shows a thin colder area between the window frame and the reveal (Fig. 7-87). According to the thermal image and with the knowledge about the planned construction, it was suggested that the thermal insulation of the reveal might be missing.

As a consequence of the expert's report, the construction company removed the exterior cover plate of the reveal and discovered that the thermal insulation had not been fit all the way to the reveal (see Fig. 7-88). The conclusion drawn from the thermography measurement was proven to be correct.

Fig. 7-86 Overview of the apartment building with brick facing: The renter complained about mold formation at the transition between the window frame and the reveal.

Date	Temperature boundary condition		Additional information
	Exterior temperature	Interior temperature	
09.01.2008 afternoon	approx. +5°C	approx. +19°C	Exterior temperature 24 hours before the measurement: +3°C to +7°C

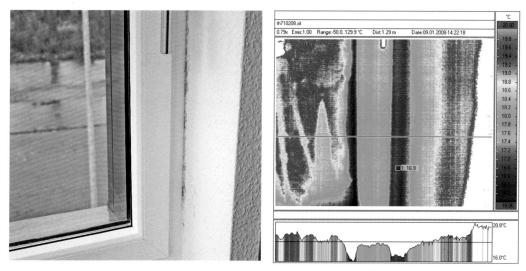

Fig. 7-87 Transition between the reveal and the window frame: the temperature distribution is remarkably cold. Note: due to the reflection of the window, the person conducting the thermography measurement often becomes visible in the thermal image.

Fig. 7-88 Uncovered exterior cover plate: (*right*) close-up of the detail, showing that the thermal insulation had not been fit properly to the reveal.

Practical Examples

7.6   Usage of Thermography in Research

7.6.1   Studies on the Applicability of the Heat Transition Coefficient (U-Value) as a Parameter for Heat Transfer Processes

In relevant standards, the heat transition coefficient U, which is determined under stationary conditions, is considered to be the most important parameter in describing the thermal properties of a construction part. Since the applicability of this parameter is the subject of controversy in some literature, a series of tests on the general thermal behavior of different construction materials was carried out at the Leibniz Universität Hannover.

For this purpose, seven different test specimens were flush-mounted in a thermally insulated hot-air oven and exposed to a predefined temperature gradient. The temperatures occurring during the heating phase and the steady-state were measured by means of temperature sensors mounted on the surfaces of the specimens. Thermography was used to visualize the heating phase and examine the selective surface temperature measuring values. To do so, the continuous shooting function that most thermography systems are equipped with was used to generate one thermal image every 60 seconds. The measuring principle used for the examination is shown in Fig. 7-89.

Subsequent to the test described above, numerical analyses and comparative calculations were carried out. In summary, the tests have shown that using the stationary U-value as a parameter to describe heat transfer processes leads to plausible and realistic results, particularly for long-term periods. In this context, thermography was used to prove the reliability of the results gained from conventional surface temperature sensors.

Detailed explanation of the tests can be reviewed in a scientific paper published in the *Bauphysik-Kalender 2005* [35].

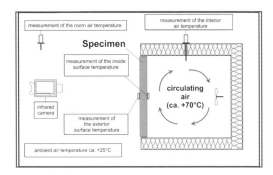 

Fig. 7-89 Measuring principle and practical test setup; the specimen being tested here is a derived timber product.

Date	Temperature boundary conditions		Additional information
	Exterior temperature	Interior temperature	
13.08.2004 morning	approx. +25°C (room temperature)	approx. +70°C (temperature in the oven)	The oven was preheated before the specimens were placed into it

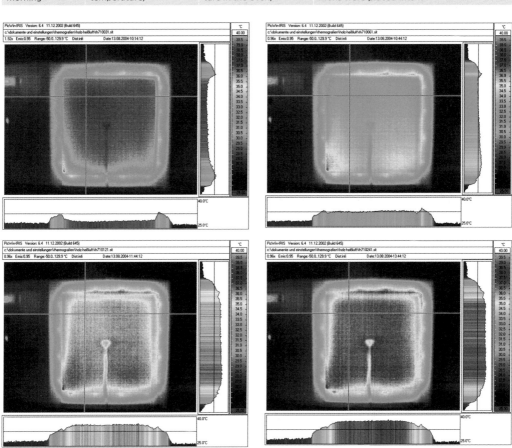

Fig. 7-90 Heating process of the derived timber specimen with proceeding test time (time span between the first and last thermal image: 3.5 hours).

7.6.2 Concrete Specimen under Cyclic Loading

In order to develop new constitutive laws for high-performance concretes under cyclic loading, (such as for wind turbines) a series of tests were conducted at the Institute for Building Materials at the Leibniz Universität Hannover. High-frequency loading and unloading are imposed on a cylindrical (hollow inside) concrete specimen by means of a high-performance hydraulic press (Fig. 7-91). The loading and unloading process causes a warming of the test specimen. Thermography can be used to localize the region within the specimen where the most heat is generated (Fig. 7-92).

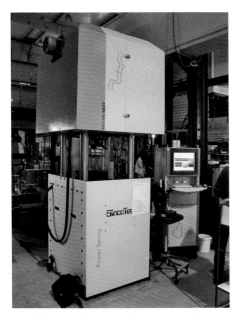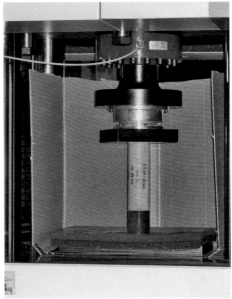

Fig. 7-91 Hydraulic press and concrete specimen.

Date	Temperature boundary conditions		Additional information
	Exterior temperature	Interior temperature	
26.01.2005 evening	./.	approx. +20°C	Test specimen in a hydraulic press under dynamic loading and unloading

Fig. 7-92 Heating of the test specimen with proceeding test time (time interval between each thermal image: ten minutes).

7.6.3　Tensile Test on a Steel Plug

Within the framework of testing metallic materials, a tensile test was performed on a steel plug (Figs. 7-93 and 7-94). As a complementary measure, the temperature distribution of the steel plug during the contraction process and failure were examined (Fig. 7-95).

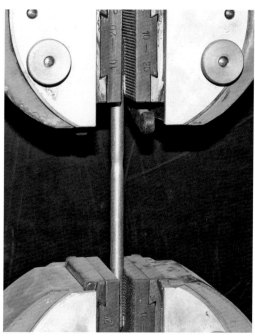

Fig. 7-93 Tensile test of a steel plug, showing contraction at the place of subsequent failure (fracture cone).

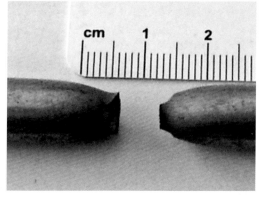

Fig. 7-94 Close-up of the fracture cone.

Date	Temperature boundary conditions		Additional information
	Exterior temperature	Interior temperature	
25.04.2005 afternoon	./.	approx. +18°C	Steel plug used for joints in timber constructions Length: 20 cm, Ø 10 mm, adjusted emissivity (ε = 0.4)

　Practical Examples

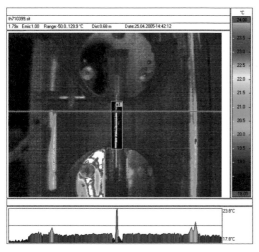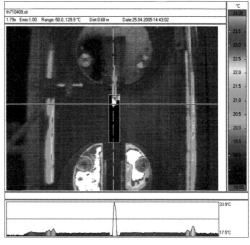

Fig. 7-95 Tensile test of a steel plug, showing contraction at the place of subsequent failure (in the form of a fracture cone): emissivity values ε in the marked area are adjusted to 0.4; the lower clamping jaws exhibit reflexion effects caused by the camera.

By means of this example, the practical meaning of the emissivity shall be pointed out: The knowledge about the correct emissivity of a certain surface is especially important when the absolute surface temperature of the measuring object is of interest.

In the example shown above, the plug is made of shiny steel. As can be taken from the header of the thermal images in Fig. 7-95, the images were taken assuming a (global) emissivity ε of 1. For the "shiny steel" an emissivity ε of 0.30 to 0.50 can be assumed to be realistic. In this case, a local adjustment of the emissivity ε to 0.40 (ambient radiation temperature θ_U = 18°C) was performed.

If the emissivity is not adjusted, the thermal image exhibits surface temperatures of the steel plug that are too low (see Fig. 7-96). By means of modern evaluation software systems, a correction of the emissivity can be performed globally (for the whole image) or locally (for a region specified beforehand).

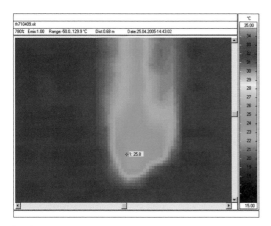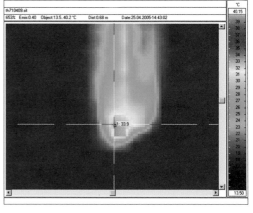

Fig. 7-96 Close-up image of the fracture cone shown in Fig. 7-95 with different emissivities: (left) ε = 1.0, displayed surface temperature θ_s = 25.8°C; (right) realistic assumption of ε = 0.4, displayed surface temperature θ_s = 33.9°C.

8 Thermographic Signatures of Typical Constructions

8.1 Wall Constructions, Building Envelope

8.1.1 Old Buildings, Partially with Internal Thermal Insulation

Constructions of old buildings exhibit a significantly higher thermal loss than new buildings or buildings refurbished in terms of thermal insulation. A typical thermal image of such a building is shown in Fig. 8-1.

Date	Temperature boundary conditions		Additional information
	Exterior temperature	Interior temperature	
02.03.2005 nighttime	approx. 1°C	approx. +22°C (estimated)	see thermal image in Fig. 8-3

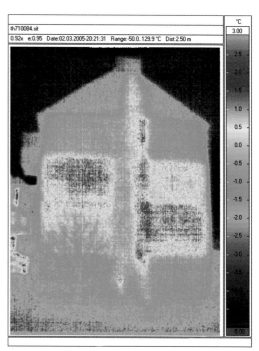

Fig. 8-1 Old building with typical checkered pattern in the thermal image.

Attic:	Undeveloped and unheated, this part of the exterior wall is significantly cooler than the other parts.
4th floor:	Based on the thermal image the conclusion could be drawn that this floor is unoccupied. Yet, according to the landlord, this floor had already been refurbished in terms of interior thermal insulation.
3rd and 2nd floors:	Occupied apartments without additional thermal insulation represent the original state of the old building.
1st floor:	Refurbished in terms of interior thermal insulation, hence, unremarkable.
Ground floor:	The thoroughfare to the backyard appears relatively warm due to its sheltered location (no emission possible).
Chimney:	The fume extraction hoods of the kitchens are connected to the chimney, which explains why the warming of the chimney begins on the first floor.

8.1.2 Old Buildings with Point Heating

In many cases, the exterior walls of (mostly) old buildings exhibit hot spots. There are two reasonable explanations for this phenomenon; the hot spots are either indicators of thermal bridge effects or, as is more often the case, they indicate the location of the heaters installed inside the building. Due to the relatively high heat demand of old buildings, the heating system has to generate a lot of heat, which due to the poor thermal insulation also results in higher thermal losses through the exterior walls. This phenomenon becomes even more striking when heaters are installed in niches, since the exterior wall is even thinner in these areas. If the heating pipes are installed within the exterior wall construction, they become apparent in the form of vertical lines (see Fig. 8-3).

Date	Temperature boundary conditions		Additional information
	Exterior temperature	Interior temperature	
14.12.2004 early morning	approx. −1°C	approx. +18°C	Temporary heating of the building two days before the on-site inspection, electrical night storage heater

 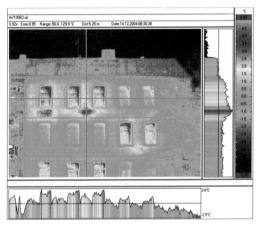

Fig. 8-2 Old building with the location of the heaters clearly standing out (here: electrical night storage heater).

Date	Temperature boundary conditions		Additional information
	Exterior temperature	Interior temperature	
02.03.2005 nighttime	approx. −1°C	approx. +24°C (estimated)	The black frame indicates the section shown in the thermal image

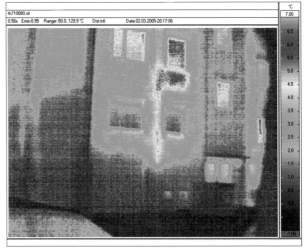

Fig. 8-3 Thermal image of an old building clearly showing the radiator and the heating pipe. On the 1st floor above the thoroughfare, the difference between the wall with additional interior thermal insulation (*left*) and the original wall with the windows (*right*) is visible.

8.1.3 Rear-Ventilated Exterior Walls

Rear-ventilated exterior wall constructions usually feature an air layer with a depth of about 4 to 6 cm between the facing and the thermal insulation layer. This air layer is connected to the outside air in the form of ventilation slots on the top and bottom edge of the facing. These slots allow for air circulation in the interlayer and are also necessary to allow for the exhaust of moisture and damp air caused by driving rain or vapor diffusion.

Due to the air circulation, the facing exhibits a nearly uniform temperature (Fig. 8-5) and hence, potential construction-related temperature differences behind the ventilation layer cannot be detected. For that reason, such constructions can only be examined by means of indoor thermography (see Chapter 7.4.6).

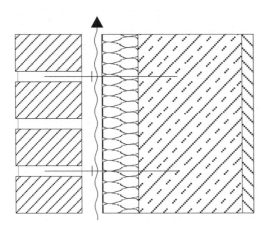

Fig. 8-4 Design layout of a rear-ventilated wall construction; the snaking line indicates the ventilation layer.

Date	Temperature boundary conditions		Additional information
	Exterior temperature	Interior temperature	
28.01.2005 morning	approx. −1°C	approx. +22°C	rear-ventilated, double-shell exterior wall construction

 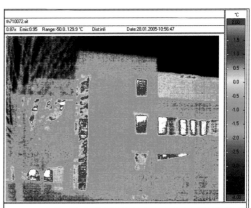

Fig. 8-5 Rear-ventilated wall construction. Due to the homogenization of the rear-ventilated facing, there are hardly any visible temperature differences.

Another example of a typical thermal image of a rear-ventilated facade free from defects is depicted in Fig. 8-6. The surface temperatures are unremarkably low. Hot spots are visible at the transition of the ventilation outlets.

Date	Temperature boundary conditions		Additional information
	Exterior temperature	Interior temperature	
09.01.2008 afternoon	approx. +5°C	approx. +19°C	Exterior temperature 24 hours before the measurement: +3°C to +7°C

Fig. 8-6 Example of a rear-ventilated cladding in the upper part of the building.

Another outdoor thermography of a rear-ventilated storage building is shown in Fig. 8-7. Here, the thermal image also shows an anomaly at a joint between the facade elements. An explanation for this could be an air leak or poor workmanship on the hidden thermal insulation.

Date	Temperature boundary conditions		Additional information
	Exterior temperature	Interior temperature	
22.12.2007 afternoon	approx. –3°C	approx. +10°C (estimated)	Overcast sky, long cold period

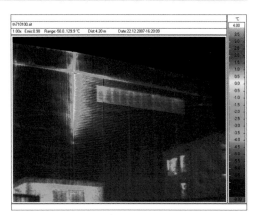

Fig. 8-7 Thermal image of a storage building. In many cases, abnormalities within rear-ventilated facades become visible by means of outdoor thermography. To determine the cause however, other examination methods are used in most cases.

8.1.4   Interior Wall Constructions with Interiors Cladding

The problem of the temperature homogenization of rear-ventilated constructions described in Section 8.1.3 can also be observed at constructions with interior cladding. The thermal images obtained from an indoor thermography do not allow for conclusions about the situation behind the cladding. In this example, the cladding is decorated with golden ornaments. For further information on how to interpret the surface temperatures of the ornaments shown in the thermal image, see the explanation on emissivity ε discussed in Chapter 2.1.2.

Date	Temperature boundary conditions		Additional information
	Exterior temperature	Interior temperature	
28.01.2005 morning	approx. –1°C	approx. +22°C	rear-ventilated double-shell exterior wall construction

Fig. 8-8 Transition between wall and ceiling of a historical building; wood cladding with ornaments.

Fig. 8-9 Wood cladding in a historical building: details of the construction concealed behind the cladding are undistinguishable.

8.1.5 Extensive Glazing

Typical examples of extensive glazing are showrooms of car dealerships. A characteristic thermogram of such a building is shown in Fig. 8-10. The elevated temperatures in the area around the post and mullion construction are remarkable. In order to determine more precise absolute temperature values, the input of the emissivity ε and the ambient temperature θ_U is necessary. Yet the indication of precise absolute temperature values of the extensive glazing shown in the picture is not that simple. In addition to the reflection properties and emissivity, the glazing's transmissibility also has to be considered for shortwave infrared systems (wavelength range of 3–5 µm).

Date	Temperature boundary conditions		Additional information
	Exterior temperature	Interior temperature	
21.11.2005 evening	approx. 0°C	approx. +20°C	./.

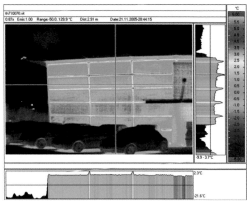

Fig. 8-10 Extensively glazed showroom: the thermogram is simplified with an emissivity ε of 1.

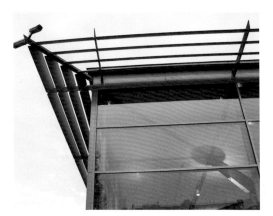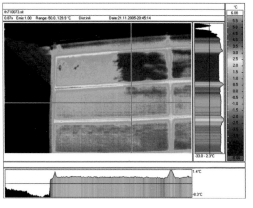

Fig. 8-11 Detail of the corner area: again, the thermogram is simplified with an emissivity ε of 1.

When thermographing glazing, it must always be taken into account that different window surface temperatures do not necessarily mean different window qualities. This can also be caused by different room temperatures, heat sources under the window, curtains in front of the windows, and the already mentioned reflections. However, thermography can provide an indication for an initial assessment; as a rule, the determination of window quality should be confirmed through further investigation (such as a window stamp, layer determination, or manufacturer's certificate).

With the abovementioned imponderables about the actual temperature on the glazing, windows that have a different structure from that of the surrounding windows are usually clearly noticeable. Subsequent installations or windows with a different function can thus be detected quickly.

Fig. 8-12 Thermogram of a glazing in a staircase, in which one window was retrofitted to be openable. This window has insulating glass,whereas the other fixed windows have only a single glazing.

Tip:

In the case of mullion/transom constructions with windows, there is sometimes the question of whether the water drainage paths in the construction are guaranteed. For this purpose, water is usually introduced into the construction and the path observed. If documentation is required, it may also be possible to use coloring substances such as uranine to mark the waterway. The disadvantage of this type of documentation is usually a long-lasting discoloring of the affected surfaces. By using warm water and the associated warming of the surface, a documentation of water drainage paths can be made with a thermogram (see Fig. 8-13).

Fig. 8-13 Documentation of water drainage paths with warm water at a window (Source: M. Krätschell, Berlin).

8.2 Other Constructions/Materials

8.2.1 Thermography of Plastics: Critical Reflection

The following thermography shows the former CargoLifter fabrication hangar, a huge self-supporting hangar construction with a length of 360 m and a height of 107 m. For its new utilization concept as Tropical Island (a swimming bath and venue with a tropical rain forest indoors) the interior temperatures of the construction are about +25°C to +28°C.

The peripheral construction at the time of the thermography measurement was primarily made of double-shell, PVC-coated polyester membranes (PES) with a top coat. Subsequently, the membrane of the middle part was replaced with a transparent and UV-transmissive ETFE-membrane.

The thermal image depicted in Fig. 8-15 is to be used as an example for the critical questioning of the interpretation of a thermogram.

Date	Temperature boundary conditions		Additional information
	Exterior temperature	Interior temperature	
26.02.2005 midday	approx. –1°C	approx. +25°C	./.

Fig. 8-14 Former CargoLifter hangar, now Tropical Island.   Fig. 8-15 Thermal image of Tropical Island.

Boundary Conditions

The measuring distance was about 700 m. During the measurement, the weather was hazy with beginning snowfall. As discussed in Chapter 2.2, radiation degrades on its way from the measuring object to the IR detector through the atmosphere. Under the circumstances of such a great measuring distance (uncommon for construction thermography) and the hazy weather, the impact of the atmosphere cannot be regarded as negligible.

Geometrical Resolution

For a geometrical resolution of the camera system of 1.58 mrad, the measuring dot of one image pixel is about 1.58 · 700 m · optical influence ≈ 1.15 m. Further detail on what led to the elevated surface temperatures in some areas of the thermal image cannot be identified with this measuring dot size. Regarding the temperatures indicated in the thermogram, refer to the distance effect explained in Chapter 7.5.2.

Measuring of Plastics

The behavior of the emissivity and transmittance (size, angle dependence, temperature dependence) of thermographed plastic foils with a top coat is not exactly known, but it has an essential impact on precise temperature determination.

In summary, the thermal image depicted in Fig. 8-15 should only be considered as an overview image that does not allow for a detailed evaluation of the construction.

8.2.2   Application of Thermography outside Civil Engineering

Thermography is not only applied in civil engineering but in almost every other technical field, including:

- mechanical engineering and process control
- firefighting (searching for people, detection of pockets of embers)
- medical engineering (veterinary or human)
- military engineering

An example of another field of use for thermography is shown in the thermal images of aircraft below.

Date	Temperature boundary conditions		Additional information
	Exterior temperature	Interior temperature	
15.06.2005 evening	approx. +25°C	./.	Picture taken at the airfield Hannover-Langenhagen EDDV, southern runway

 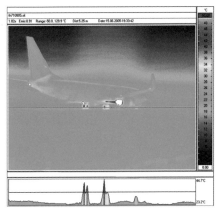

Fig. 8-16 Thermal image of a Boeing 737-800.

The area around the main landing gear of the aircraft shown above exhibits elevated temperatures due to the braking maneuver. Moreover, it can be seen that the temperature of the engine exhaust is beyond the present measuring range of the thermography camera (max. +120°C). For this field of application, the temperature range of the camera would have to be changed.

Another aircraft is shown in a thermal image in Fig. 8-17. At the aircraft body and the extended wing flaps, the effect of the angle dependence of emissivity on the temperature displayed in the image is visible. As in the thermal image shown above, the brakes of the main landing gear are heated. Another hot spot, which is caused by an exhaust opening of the aircraft's air-conditioning system, can be seen at the transition between the aircraft body and the wing.

Fig. 8-17 Thermal image of a Tupolev 154-M.

9 Summary and Short Guide into Thermogram Generation

The preceding chapters and examples have explained the theory of the infrared technology and have shown the different camera systems and the various fields of application. In this chapter, the influences that have to be considered when performing a thermography measurement are briefly summarized in note form. The instructions given in this chapter are not absolute, which means that depending on the application and the requirements the definitive criteria can vary.

9.1   General Preconditions

Personnel Requirements

- Personnel need to have extensive experience in the field of (building) physics, metrology, and building technology in general in order to be able to reliably carry out measurements by means of infrared systems.

Thermography System Requirements

- The system in use must be suitable for performing construction thermography with respect to the thermal, geometrical, and temporal resolution, the temperature range, and the sensitivity of the detector for the expected measuring temperatures.
- The evaluation software must be suitable for the postprocessing of the thermal images and the generation of a meaningful report of the conducted examination.

9.2   Performing the Thermographic Analysis

Meteorological Boundary Conditions

- The temperature differences between indoors and outdoors must be sufficient, and temperature fluctuations should be marginal, especially right before the thermography measurement.
- Outdoor thermography measurement should be conducted before sunrise or without the impact of direct solar irradiation.
- Outdoor thermography results can be falsified by high wind velocities (convective heat transmission). Thus, if necessary, wait for calm weather.
- Outdoor thermography is not meaningful for wet wall surfaces.

Preparation for Thermographic Examinations

- Conduct an on-site inspection before the measurement; identify the existing construction from construction drawings, specifications, or test drillings.
- The building should be heated to a steady and sufficient temperature prior to the thermography measurement; open interior doors to allow for uniform heating. If necessary, move furniture and fixtures prior to the heating.

- If possible, record the indoor climate by means of data loggers.
- Provide the owner, renter, and neighbors with general information about the nighttime thermography to avoid false alarms (like calling the police). Determine the measuring sites beforehand to assure safe access in the dark.
- Take reference photos of the area that will be depicted in the thermal image (these photos can be taken before, during, or after the thermography measurement at first daylight).

Impact of the Measuring Environment

- Keep the measuring distance as short as possible (compare the impact of the atmosphere and measuring dot size).
- No thermal images should be taken in fog, rain, or snow, as they are meaningless because of the measurement noise caused by effects of refraction and diffraction.
- Estimate the influence of the reflecting ambient radiation and, if necessary, measure the ambient radiation temperature.

Impact of the Measuring Object (Surface)

- The emissivity of the measuring object surface should be known. If it is only estimated, absolute temperature values in the thermal image should be questioned with respect to their accuracy.
- The surfaces depicted in the thermal image should exhibit the same – or at least a similar – emissivity. Otherwise, it should be indicated that the temperatures in regions with diverging emissivity are not correct (take, for example, framed mirrored glass in combination with wall constructions – in most cases, the emissivity of the mirrored glass is lower than that of the wall surface).
- The impact of reflections on surfaces must be considered (for example, when glazing is thermographed).
- If possible, the ambient radiation temperature should be documented (in most cases the ambient radiation temperature is recorded on the thermal image).

Recording Significant Influencing Variables

In order to generate a thermographic report and for the post processing of the thermal images, the metrological registration of certain boundary conditions is unavoidable:

- date and time of the measurement
- measuring distance between the measured object and the IR camera (to determine potential influences of the passage through the atmosphere)
- exterior air temperature at the time of the measurement
- development of the exterior temperature until about 24 hours before the measurement (in order to evaluate the general climate situation)
- relative humidity in the measuring environment (to identify potential influences of the passage through the atmosphere)

- interior temperature (in order to evaluate the general climate situation)
- wind velocity (influence of convective transmission)
- (optional) determination of surface temperatures using common measuring methods to get a reference value for the thermal images (difficulty of determining the correct emissivity, issue with the reflection of the ambient or background radiation)
- estimation or determination of the background radiation temperature θ_U. When conducting outdoor thermography measurements, the climate boundary conditions may provide information (for example, overcast sky: radiation temperature is similar to the air temperature; clear sky: mostly, radiation temperature is significantly lower than the air temperature).

Writing an Examination Report

The form and the content of an examination report generally depends on the problem. In general, the inclusion of the following information is useful:

- purpose and goal of the thermography measurement and description of the measuring object (for example, structure of the construction, light/heavy construction, etc.)
- measuring time, climate data, particularities
- information about the thermographic technology and software used
- thermal images in combination with elevation, floor plans, and/or photos of the building
- explanations on the thermal images, evaluation, indication of surface temperatures or temperature differences (respecting the accuracy of the absolute temperature values)
- conclusions regarding the defined problem.

10 Appendix

10.1 Nomenclature

Parameters

α	[./.]	absorption coefficient
α_s	[./.]	absorption coefficient of a black body, $_s$ = 1
X	[W/K]	coefficient for determining the losses due to concentrated thermal bridges
ε	[./.]	emissivity
ε_M	[./.]	emissivity of the measuring object
ε_{MP}	[./.]	emissivity of the measuring path
λ	[μm]	wavelength
λ_{max}	[μm]	maximum radiance
λ	[W/(mK)]	thermal conductivity
θ_e	[°C]	exterior temperature
θ_{HS}	[K]	background radiation temperature
θ_i	[°C]	interior temperature
θ_M	[K]	(radiation) temperature of the measuring object
θ_{MP}	[K]	radiation temperature of the measuring path
θ_{si}	[°C]	surface temperature interior (s = surface)
θ_{se}	[°C]	surface temperature exterior (s = surface)
$\theta_{si,cr}$	[°C]	critical interior surface temperature to prevent mold formation
θ_U	[K]	ambient radiation temperature (average temperature of the half-space in front of or over the measuring object)
σ	[W/(m²K⁴)]	Stefan-Boltzmann coefficient = $5.67 \cdot 10^{-8}$ W/(m²K⁴)
ρ	[./.]	reflectance
ρ_M	[./.]	reflectance of the measuring object
ρ_{MP}	[./.]	reflectance of the measuring path
τ	[./.]	transmittance
τ_M	[./.]	transmittance of the measuring object
τ_{MP}	[./.]	transmittance of the atmosphere (for the infrared radiation passing through the atmosphere on its measuring path)
ψ	[W/(mK)]	thermal bridge loss coefficient accounting for linear thermal bridges (for use in EnEV, "energy saving rule" in Germany)
A	[m²]	area
A	[W/m²]	generally: absorbed radiation
c_1	[W · cm⁻² · μm⁴]	1st radiation coefficient = $3.7418 \cdot 10^4$
c_2	[K · μm]	2nd radiation coefficient = $1.4388 \cdot 10^4$ [K · μm]
C_s	[W/(m²K⁴)]	radiation coefficient of a black body = $5.67 \cdot 10^{-8}$ W/(m²K⁴)
f_{Rsi}	[./.]	temperature factor for the computational evaluation of thermal bridges according to DIN 4108-2
h_{se}	[W/(m²K)]	heat transfer coefficient, exterior
h_{si}	[W/(m²K)]	heat transfer coefficient, interior
H_s	[W/m²]	background radiation, $f(\theta_{HS})$
H_T	[W/K]	transmission heat loss (EnEV)
M_λ	[W/(cm² · μm)]	spectral specific radiation of a black body

M_Σ	[W/m²]	radiation, measured directly at a real body, without the influence of the measuring path
M'_Σ	[W/m²]	measurable radiation of a real body, degraded by the influence of the measuring path
M	[W/m²]	specific radiation of a black body or in general of the incident radiation
M_M	[W/m²]	emitted radiation of a real body, $f(\theta_M)$
M_{MP}	[W/m²]	radiation emitted by the measuring path, $f(\theta_{MP})$
M_{real}	[W/m²]	specific radiation of a real body
$M_{black\,body}$	[W/m²]	specific radiation of a black body
M_{IR}	[W/m²]	radiation arriving at the infrared detector
R	[W/m²]	generally: reflected radiation
R_{se}	[m²K/W]	heat transmission resistance, exterior
R_{si}	[m²K/W]	heat transmission resistance, interior
R_U	[W/m²]	reflected ambient radiation
R'_w	[dB]	weighted airborne sound reduction index
T	[K]	absolute temperature
T	[W/m²]	generally: transmitted radiation
T_{HS}	[W/m²]	transmitted background radiation
U	[W/m²]	ambient radiation, $f(\theta_U)$
U	[W/(m²K)]	thermal transmission coefficient

Abbreviations

ETFE	[./.]	**E**thylene-**T**etra**F**luor**E**thylene – Synthetics used for foils, particularly good transmittance and UV-transmittance
FOV	[°]	**F**ield **O**f **V**iew
FPA	[./.]	**F**ocal-**P**lane-**A**rray – Detector type with single detectors arranged in matrix form
IFOV	[mrad]	**I**nstantaneous **F**ield **O**f **V**iew (Field of view of a single detector)
NETD	[K]	**N**oise **E**quivalent **T**emperature **D**ifference
NUC	[./.]	**N**on **U**niformity **C**orrection for the compensation of the reciprocal influence of neighboring detectors
PES	[./.]	**P**oly**E**ther**S**ulfon – Synthetics (Fiber material for backing fabric)
PVC	[./.]	**P**oly**V**inyl**C**hloride – Thermoplastic synthetics
WDVS	[./.]	**W**ärme**D**ämm**V**erbund-**S**ystem (engl.: External Thermal Insulation Composite Systems (ETICS))

10.2 Bibliography

[1]   ASTM C 1046-95(2013): Standard Practice for In-Situ Measurement of Heat Flux and Temperature on Building Envelope Components. West Conshohocken, PA: ASTM International, 2013.

[2]   ASTM C 1060-11a: Standard Practice for Thermographic Inspection of Insulation Installations in Envelope Cavities of Frame Buildings. West Conshohocken, PA: ASTM International, 2011.

[3]   ASTM D 4788-03(2013): Standard Test Method for Detecting Delaminations in Bridge Decks Using Infrared Thermography. West Conshohocken, PA: ASTM International, 2013.

[4]   Baehr, Hans Dieter, and Karl Stephan. *Wärme- und Stoffübertragung*, 4th ed. Berlin: Springer-Verlag, 2004.

[5]   Tanner, Christoph, Beat Lehmann, and Thomas Frank. "Energetische Beurteilung von Gebäuden mit Thermografie und der Methode QualiThermo," final report for the research project "Energetische Beurteilung von Gebäuden mittels Infrarotbildern (Quali-Thermo)," Eidgenössisches Departement für Umwelt, Verkehr, Energie und Kommunikation UVEK, https://www.thech.ch/de/thermografie/bauthermografie/qualithermo.

[6]   Krätschell, Michael, and Frank Anders. *Schäden durch mangelhaften Wärmeschutz*, 2nd rev. and exp. ed., Reihe Schadenfreies Bauen 32. Stuttgart: Fraunhofer IRB, 2012.

[7]   Bundesministerium für Raumordnung, Bauwesen und Städtebau. *Dritter Bericht über Schäden an Gebäuden*. Bonn, 1995.

[8]   Deutscher Wetterdienst DWD. Handbuch zu den neuen Testreferenzjahren 2011 (TRY). Offenbach: Deutscher Wetterdienst DWD, 2011, http://www.irbnet.de/daten/baufo/20118035316/TRY_Handbuch.pdf.

[9]   DELPHIN, Simulationsprogramm für den gekoppelten Wärme-, Luft-, Feuchte-, Schadstoff- und Salztransport, Institut für Bauklimatik, TU Dresden.

[10] DGZfP Merkblatt TH-1: Charakterisierung von Thermographiesystemen. Berlin: DGZfP, 1999.

[11] DGZfP Merkblatt B 05: Merkblatt über das aktive Thermographieverfahren zur Zerstörungsfreien Prüfung im Bauwesen. Berlin: DGZfP, 2013.

[12] DIN 4108-2:2013-02: Wärmeschutz und Energie-Einsparung in Gebäuden – Teil 2: Mindestanforderungen an den Wärmeschutz (Thermal protection and energy economy in buildings – Part 2: Minimum requirements to thermal insulation). Berlin: Beuth, 2013.

[13] DIN 5031: Strahlungsphysik im optischen Bereich und Lichttechnik (Optical radiation physics and illuminating engineering), parts 1–11. Berlin: Beuth, 1982–2018.

[14] DIN EN 13187:1999-05: Wärmetechnisches Verhalten von Gebäuden, Nachweis von Wärmebrücken in Gebäudehüllen, Infrarot-Verfahren (Thermal performance of buildings – Qualitative detection of thermal irregularities in building envelopes – Infrared method). Berlin: Beuth, 1999.

[15] DIN EN ISO 13788:2013-05: Wärme- und feuchtetechnisches Verhalten von Bauteilen und Bauelementen – Raumseitige Oberflächentemperatur zur Vermeidung kritischer Oberflächenfeuchte und Tauwasserbildung im Bauteilinneren – Berechnungsverfahren (ISO 13788:2012); German version EN ISO 13788:2012 (Hygrothermal performance of building components and building elements – Internal surface temperature to avoid critical surface humidity and interstitial condensation – Calculation methods). Berlin: Beuth, 2013.

[16] DIN EN ISO 9712:2012-12: Zerstörungsfreie Prüfung – Qualifizierung und Zertifizierung von Personal der zerstörungsfreien Prüfung (Non-destructive testing – Qualification and certification of NDT personnel [ISO 9712:2012]). Berlin: Beuth, 2012.

[17] DIN EN ISO 6781-3:2016-05: Verhalten von Gebäuden – Feststellung von wärme- luft- und feuchtebezogenen Unregelmäßigkeiten in Gebäuden durch Infrarotverfahren – Teil 3: Qualifikation der Ausrüstungsbetreiber, Datenanalytiker und Berichtsautoren (Performance of buildings – Detection of heat, air and moisture irregularities in buildings by infrared methods – Part 3: Qualifications of equipment operators, data analysts and report writers [ISO 6781-3:2015]). Berlin: Beuth, 2016.

[18] DIN EN 16714-1:2016-11: Zerstörungsfreie Prüfung – Thermografische Prüfung – Teil 1: Allgemeine Grundlagen; German version, EN 16714-1:2016 (Non-destructive testing – Thermographic testing – Part 1: General principles)

[19] DIN EN 16714-2:2016-11: Zerstörungsfreie Prüfung – Thermografische Prüfung – Teil 2: Geräte, German version, EN 16714-2:2016 (Non-destructive testing – Thermographic testing – Part 2: Equipment). Berlin: Beuth, 2016.

[20] DIN EN 16417-3:2016-11: Zerstörungsfreie Prüfung – Thermografische Prüfung – Teil 3: Begriffe, German version, EN 16714-3:2016 (Non-destructive testing – Thermographic testing – Part 3: Terms and definitions). Berlin: Beuth, 2016.

[21] DIN EN 17119:2018-10: Zerstörungsfreie Prüfung – Thermografische Prüfung – Aktive Thermografie, German version, EN 17119:2018 (Non-destructive testing – Thermographic testing – Active thermography). Berlin: Beuth, 2018.

[22] DIN EN ISO 6946:2018-03: Bauteile – Wärmedurchlasswiderstand und Wärmedurchgangskoeffizient – Berechnungsverfahren (Building components and building elements – Thermal resistance and thermal transmittance – Calculation methods); German version, EN ISO 6946:2017. Berlin: Beuth, 2018.

[23] Dittié, Georg. "Die Aussagekraft von Thermogrammen," lecture given at the seminar *Thermografie am Bau*, DGZfP (Deutsche Gesellschaft für Zerstörungsfreie Prüfung e.V.), Berlin, April 28, 2009.

[24] Fischer, Nico. "Einflüsse von klimatischen und baulichen Randbedingungen auf die Detektierbarkeit von Unregelmäßigkeiten in der Gebäudethermografie." Diplomarbeit, Institut für Bauphysik, Leibniz Universität Hannover, 2011.

[25] Gubareff, G. G., J. E. Janssen, and T. H. Torborg. *Thermal Radiation Properties Survey: A Review of the Literature*, 2nd ed. Minneapolis, MN: Honeywell Research Center, 1960.

[26] Hauser, Gerd. "Wärmebrücken," in *Bauphysik-Kalender 2001*, ed. Erich Cziesielski. Berlin: Ernst & Sohn, 2001.

[27] ISO 6781:1983: Thermal insulation – Qualitative detection of thermal irregularities in building envelopes – Infrared method. Geneva: International Organization for Standardization, 1983.

[28] Jenisch, Richard, and Martin Stohrer. *Tauwasserschäden*, 2nd ed. Schadenfreies Bauen 16. Stuttgart: Fraunhofer IRB, 2001.

[29] Künzel, Hartwig M. "Verfahren zur ein- und zweidimensionalen Berechnung des gekoppelten Wärme- und Feuchtetransports in Bauteilen mit einfachen Kennwerten." Dissertation, Universität Stuttgart, 1994.

[30] Raicu, Alexandra. *IR-Thermografie im Bauwesen: Aufstellung eines Leitfadens zur Anwendung der Infrarotthermografie bei instationären Temperaturverhältnissen zur Feststellung versteckter Baufehler*. Stuttgart: Fraunhofer IRB, 2000.

[31] NEC San-ei Instruments, Ltd. Thermo Tracer Series TH 7102, Operation Manual, June 2002.

[32] Nehring, Gerhard. "Über den Wärmefluß durch Außenwände und Dächer in klimatisierte Räume infolge der periodischen Tagesgänge der bestimmenden meteorologischen Elemente," *Gesundheits-Ingenieur* 83, no. 7 (1962): 185–89; no. 8 (1962): 230–42; and no. 9 (1962): 253–69.

[33] Flixo software system for the calculation of two-dimensional thermal bridges, Infomind GmbH, Thun, Switzerland.

[34] Rahn, Axel. "Sanierung von Wärmebrücken durch aktive und passive Beheizung," in *Ingenieur-Hochbau Berichte aus Forshung und Praxis: Festschrift zum 60. Geburtstag von Prof. Dr. Erich Cziesielski*. Düsseldorf: Werner, 1998.

[35] Richter, Torsten, and Steffi Winkelmann-Fouad. "Anwendung des U-Wertes als Kenngröße für Wärmetransportvorgänge," in *Bauphysik-Kalender 2005*, ed. Erich Cziesielski. Berlin: Ernst & Sohn, 2005.

[36] Schuster, Kolobrodov. *Infrarotthermographie*, 2nd ed. Weinheim: Wiley-VCH Verlag, 2004.

[37] Sedlbauer, Klaus. "Vorhersage von Schimmelpilzbildung auf und in Bauteilen." Dissertation, Universität Stuttgart, 2001.

[38] Testo AG. *Pocket-Guide Thermografie, Theorie – Praxis – Tipps & Tricks*. Lenzkirch: Testo AG, 2008.

[39] Tingwaldt, D. P., and H. Kunz. "Optische Temperaturmessung," in *Zahlenwerk und Funktionen*, 6th ed., IV/4a Technik, 47–147. Berlin: Springer Verlag, 1967.

[40] VATh (Verband für angewandte Thermografie e.V.). VATh-Richtlinie: Bauthermografie, June 2016.

[41] VDI/VDE 3511 Blatt 4: 2011-12: Technische Temperaturmessungen, Strahlungsthermometrie (Temperature measurement in industry – Radiation thermometry). Berlin: Deutsches Institut für Normung, 2011.

[42] Walther, Andrei, Christiane Maierhofer, Mathias Röllig, Frank U. Vogdt, and Manuela Walsdorf-Maul. "Die aktive Thermografie: Ein Beitrag zur Qualitätssicherung im Bauwesen," paper presented at the conference *Thermografiekolloquium 2009*, Stuttgart, October 8–9, 2009.

[43] Walther, Andrei, Christiane Maierhofer, Bernd Hillemeier, Carsten Rieck, Mathias Röllig, Ralf Arndt, and Horst Scheel. "Zerstörungsfreie Ortung von Fehlstellen und Inhomogenitäten in Bauteilen mit der Impuls-Thermografie," *Beitrag in Bautechnik* 81, no. 10 (2004): 786–93.

[44] Wiggenhauser, Herbert, Alexander Taffe, et al. "Zerstörungsfreie Prüfung im Bauwesen," in *Bauphysik-Kalender 2004*, ed. Erich Cziesielski. Berlin: Ernst & Sohn, 2004.

[45] WUFI-Software for the calculation of coupled heat and moisture transport in building components; versions include WUFI Pro, WUFI PLUS, and WUFI 2D. Stuttgart: Fraunhofer-Institut für Bauphysik IBP.

[46] Software WUFI-Bio, which assesses the risk of mold growth. Stuttgart, Fraunhofer-Institut für Bauphysik IBP.

[47] WTA (Wissenschaftlich-Technischer Arbeitsgemeinschaft für Bauwerkserhaltung und Denkmalpflege e.V.).
Merkblatt 6-1-01/D (version May 2002): Leitfaden für hygrothermische Simulationsberechnungen (A guide to hygrothermal computer simulation). Stuttgart: Fraunhofer IRB, 2002.
Merkblatt 6-2 (version December 2014): Simulation wärme- und feuchtetechnischer Prozesse (Simulation of heat and moisture transfer). Stuttgart: Fraunhofer IRB, 2014.
Merkblatt 6-3-05/D (version April 2006): Rechnerische Prognose des Schimmelpilzwachstumsrisikos (Calculative prognosis of mould growth risk). Stuttgart: Fraunhofer IRB, 2006.

[48] WTA (Wissenschaftlich-Technischer Arbeitsgemeinschaft für Bauwerkserhaltung und Denkmalpflege e.V.) Merkblatt zur Thermografie:
Merkblatt 6-18-19/D:2019-02: Bauthermografie im Bestand. Stuttgart: Fraunhofer IRB, 2019.

[49] EnEV 2014, Energieeinsparverordnung (energy saving regulation): Second amendment to the Energieeinsparverordnung of November 18, 2013, published in the Bundes-gesetzblatt November 21, 2013, legally effective May 1, 2014, to be replaced by the GEG Gebäude Energie Gesetz (building energy law) approved on October 23, 2019.

[50] Rödel, Andreas. "Leckageortung an Bauwerksabdichtungen," in *Bauphysik-Kalender 2002*, ed. Erich Cziesielski. Berlin: Ernst & Sohn, 2002.

10.3   Acknowledgments

We would like to thank the following persons for their cooperation in translating the English edition, creating the illustrations, suggestions for relevant examples and much more:

Heide Ackerbauer, Nico Feige, Nico Fischer, the Fuhlbrück family, Naghmeh Hajibeik, the Helmke family, the Kluge family, Tomek Kniola, Michael Krätschell, Adrian Kreißig, the family of Doris and Bodo Richter, the family of Sandy and Frank Richter, Tim Thiemann, Andrei Walther, and Steffi Winkelmann-Fouad.

Thank you!

10.4 Index

10.5 Autoren

Dr.-Ing. Torsten Richter, born in 1972, vocational training as a skilled construction worker with additional high school diploma, studied civil engineering at the Technical University of Berlin, graduated in 1999, doctorate in 2009, senior engineer at the Institute of Building Physics at Leibniz University in Hannover. Since 2011, publicly appointed and accredited expert for moisture protection and thermal insulation with a focus on building thermography; since 2000, well-founded practical experience as an employee in renowned engineering offices in the entire field of building physics; since 2017, partner in own engineering office in the field of building physics.

Univ.-Prof. Dr.-Ing. Nabil A. Fouad, born in 1964, studied civil engineering at the Ain Shams University in Cairo, graduated 1986, doctorate in 1997 from the Technical University of Berlin. Professor of building planning and building maintenance at the Leibniz University in Hannover in 2001, and since 2007 professor of building physics and building maintenance at the same university. Managing partner of an engineering office for building planning and building mainte-nance. Numerous publications and research work, member of standards and expert committees as well as publicly appointed and accredited expert with the tenor "Building Physics and Fire Protection."

Copy editing: Keonaona Petersen
Project management: Alexander Felix, Nora Kempkens
Cover design: Harald Pridgar
Layout: Georgia Zechlin, adapted by Harald Pridgar
Typesetting: Sven Schrape
Production: Amelie Solbrig
Paper: Magno Volume, 135 g/m^2
Printing: Beltz Grafische Betriebe GmbH, Bad Langensalza
Prepress: LVD Gesellschaft für Datenverarbeitung mbH, Berlin

Library of Congress Control Number: 2021938761

Bibliographic information published by the German
National Library
The German National Library lists this publication in the
Deutsche Nationalbibliografie; detailed bibliographic
data are available on the Internet at http://dnb.dnb.de.

ISBN 978-3-0356-2267-6

e-ISBN (PDF) 978-3-0356-2268-3

Expanded and updated English license edition based
on the original German edition: © 2012, *Leitfaden
Thermografie im Bauwesen* by Fraunhofer IRB Verlag,
ISBN (Print): 978-3-8167-8456-2

© 2021 Birkhäuser Verlag GmbH, Basel
P.O. Box 44, 4009 Basel, Switzerland
Part of Walter de Gruyter GmbH, Berlin/Boston

9 8 7 6 5 4 3 2 1
www.birkhauser.com